Additive Manufacturing and Industry 4.0

The text covers four important areas: digital manufacturing, modern manufacturing processes, modeling and simulation in smart industry, and nanotechnology. It further presents mathematical models to represent physical phenomena and applies modern computing methods and simulations in analyzing them. The text covers key concepts such as abrasive flow machining (AFM), abrasive water jet (AWJ) machining, and hybrid machining for micro-/nanomanufacturing. It will serve as an ideal reference text for senior undergraduate, graduate students, and researchers in fields including mechanical engineering, aerospace engineering, manufacturing engineering, and production engineering.

Features

- Discusses the sustainable development aspects of additive manufacturing in Industry 4.0.
- Studies electrochemical machining processes for micromachining.
- Presents experimental investigation of friction factor and heat transfer rate in the laminar regime.
- Examines the mechanical and microstructural characterization of titanium chips using large strain machining.
- Covers hybrid approaches like electrochemical machining and magnetic abrasive flow machining.

The book emphasizes linking the computer interface with the digital manufacturing process and their demonstration using commercially available software like Solid-Edge, ProE, and CATIA. It further discusses important aspects of digital manufacturing, advanced composites, artificial intelligence, and modern manufacturing processes.

Advances in Manufacturing, Design and Computational Intelligence Techniques

Series Editor: Ashwani Kumar

Advanced Materials for Biomedical Applications
Ashwani Kumar, Yatika Gori, Avinash Kumar, Chandan Swaroop Meena and Nitesh Dutt

Additive Manufacturing in Industry 4.0
Methods, Techniques, Modeling, and Nano Aspects
Vipin Kumar Sharma, Ashwani Kumar, Manoj Gupta, Vinod Kumar, Dinesh Kumar Sharma, and Subodh Kumar Sharma

For more information about this series, please visit:
www.routledge.com/Advances-In-Manufacturing-Design-and-Computational-Intelligence-Techniques/book-series/%20CRCAIMDCIT

Additive Manufacturing in Industry 4.0

Methods, Techniques, Modeling, and Nano Aspects

Edited by
Vipin Kumar Sharma, Ashwani Kumar,
Manoj Gupta, Vinod Kumar, Dinesh Kumar Sharma,
and Subodh Kumar Sharma

CRC Press
Taylor & Francis Group
Boca Raton London New York

CRC Press is an imprint of the
Taylor & Francis Group, an **informa** business

First edition published 2023
by CRC Press
6000 Broken Sound Parkway NW, Suite 300, Boca Raton, FL 33487-2742

and by CRC Press
4 Park Square, Milton Park, Abingdon, Oxon, OX14 4RN

CRC Press is an imprint of Taylor & Francis Group, LLC

ISBN: 9781032392844 (hbk)
ISBN: 9781032418414 (pbk)
ISBN: 9781003360001 (ebk)

DOI: 10.1201/9781003360001

Typeset in Sabon
by Newgen Publishing UK

This book is dedicated to all engineers, researchers, and academicians . . .

Contents

Aim and scope

In the era of globalization, the approach of every industry shifted toward additive manufacturing for the optimum utilization of input resources, i.e. man, machine, and materials, so that better productivity might be achieved by way of rapid prototyping, 3-D printing, and digital manufacturing. This book series is primarily designed for the fundamental understanding of the additive manufacturing approach in core sector industries related to mechanical, electronic, civil, computer applications and the biomechanics research fraternity. It is intended to fill the gap of knowledge regarding additive manufacturing in academia and industry.

This book series is a result of the collective and multidisciplinary efforts of persons around a common theme: *Additive Manufacturing in Industry 4.0: Methods, Techniques, Modeling, and Nano Aspects*. This book is based on the original idea of additive manufacturing, i.e. connecting high-quality production efficiency with pioneering innovation and sustainable practices adopted by industries for the betterment of society around the world. In connection with the original idea of additive manufacturing, a vision of creating digital manufacturing is put forth as a future way for industries to maintain sustainability in a competitive world. The aim of the book series reflects the technological advancements in industry along with the related positive societal changes. Hopefully, this book series opens new vistas for academicians, industrialists, researchers, and scientist working in additive manufacturing.

The series is also very beneficial for research scholars or postgraduates working in the different domains of artificial intelligence, the Internet of Things, digital manufacturing, modern manufacturing processes, advanced composites, and other such fields. The content of the book is purely related to recent trends in additive manufacturing in various segments of the manufacturing sector.

The book series is designed in such a way that it will cover all aspects of basic and applied research applications in the smart factory system. Therefore, both academicians and industrial practitioners will benefit from this book series. Special emphasis is given to the computer interface with the digital manufacturing process and to the demonstration of using commercially available software like Solid-Edge, ProE, and CATIA. The series covers almost all directions taken by additive manufacturing technologies, blending fundamental development with recent technologies. The audience will attain a fundamental understanding of different perspectives and recent developments after a careful reading of book content, reinforced by deep study. This book series is presented in a lucid and simplified way to make it enjoyable for newcomers to the field. The theme of the series revolves around four modules: digital manufacturing, modern manufacturing, modeling and simulation in smart industry, and nanotechnology.

Digital manufacturing: The first module of the book series will cover topics related to newer trends in advanced manufacturing by the means of a computer-aided approach in digital manufacturing. Research areas of interest in computer-aided digital manufacturing technologies cover new trends, including value-adding manufacturing through IT-based tools. The main areas of focus include innovative design methodologies in mechanical engineering, biomedical devices printing, multimaterials modeling, design and manufacturing, design and manufacturing of structural or functional components, sustainable design/manufacturing, artificial intelligence applications in manufacturing, virtual reality, and the impact of computer-aided digital manufacturing technologies on the product development cycle.

Modern manufacturing: The second module will cover modern manufacturing techniques in order to achieve better productivity in industry and fulfill the Industry 4.0 approach. Research areas of interest in modern manufacturing cover the development of novel nontraditional processes in new dimensions and in combination with already existing processes with the idea of achieving new process capabilities for the needs of micro- and miniature products. The main topics include abrasive flow machining (AFM) and abrasive water jet (AWJ) machining, hybrid machining for micro-/nanomanufacturing, Equipment development for nontraditional or hybrid micro-/nanomanufacturing, vibration-assisted methods for micro-/nanomachining, micro-EDM and micro-ECM processes, micro-/ nanosurface texturing and characterization, and similar topics.

Modeling and Simulation: The third module of the book series will cover the development of mathematical models representing physical phenomena and applying modern computing methods and simulations to analyze them. The main topics include recent advances in the fields of nanomechanics and biomechanics, simulations of multiscale and multiphysics problems, developments in solid mechanics and finite element method, advancements in computational fluid dynamics and transport phenomena refrigeration plants, heat exchangers, heat pumps, and heat pipes, combined heat and power and advanced alternative cycles, combustion processes, heat transfer, solar cells, solar thermal power plants, and the integration of renewable energy with conventional processes. It also covers various hot topics like optimized design solutions with AI machine learning techniques, Big Data analytics, technological development and cloud-based computation for faster simulations used in manufacturing industries.

Nanotechnology: The nanotechnology module will be focused on functional nanomaterials, including sensors and devices; nanomaterials and films used to enhance hydrophobicity (water repellency); antireflective coatings; self-cleaning coatings; antifogging and ultraviolet resistance; nanoscale materials including rare earth oxides used to enhance mechanical strength in sports equipment like baseball bats, tennis racquets, and golf clubs, vehicle parts; and aircraft parts. Nanoscale materials are currently used in drug delivery devices, including dendrimers (nanomolecules that enable targeted drug delivery); in air and water filtration devices; in electronic devices, including transistors, nanowires, semiconducting nanotubes, and quantum processors; in alternate energy applications, including solar cells made with nanorods created by atomic layer deposition and fuel cells made with nanopolymers.

Editors
Dr. Vipin Kumar Sharma
Dr. Ashwani Kumar
Prof. (Dr.) Manoj Gupta
Dr. Vinod Kumar
Dr. Dinesh Kumar Sharma
Dr. Subodh Kumar Sharma

Preface

This new book series titled *Additive Manufacturing in Industry 4.0: Methods, Techniques, Modeling, and Nano Aspects* relates to additive manufacturing processes and their importance in technological advancement in the current era of the twenty-first century. Additive manufacturing processes form a core research area for a majority of researchers, academicians, and students. In a particular, this is an essential emerging area for all interdisciplinarians. With this in mind, an attempt has been made to include chapters related to the latest advances in additive manufacturing processes and their usefulness to the engineering fraternity.

Being a lead editor, I am indebted to all the contributors who made their contributions in the form of chapters, as well as to many others. I have spent thousands of hours identifying pertinent technology, finding leading authorities in their fields, and working with them to develop this book series. They, in turn, have also spent thousands of hours in their fields to bring these chapters to you, the reader, in both a convenient and an authoritative manner.

I encourage anyone who finds errors or omissions in this book series to contact me. Similarly, as new processes come online, I would love to hear about them; I will find a way to include them in future editions of this book. Last, editing any publication is a work of love that expands the editor's knowledge immensely; I sincerely hope that the resulting book greatly expands each reader's knowledge too.

Heartfelt thanks are due to Prof. Prakash Gopalan (Director, Thapar Institute of Engineering & Technology, Patiala), Dr. Ajay Batish (Deputy Director, Dean of Partnerships & Accreditation and Professor in Mechanical Engineering Department, Thapar Institute of Engineering & Technology, Patiala), Dr. Tarun Bera (Prof. & Head, Department of Mechanical Engineering, Thapar Institute of Engineering & Technology, Patiala), Dr. Ravinder Singh Joshi (Assistant Professor, Department of Mechanical Engineering, Thapar Institute of Engineering & Technology, Patiala) for their encouragement and support during making of this book series. I wish to thank my seniors, colleagues, and dear friends – Dr. Rajeev Kumar Agarwal, Dr. Dinesh Kumar, Dr. BN Tripathi, Dr. Pardeep Kumar, Dr. Ansar Ali SK, Mr. Sandeep Kumar, Mr. Dinesh Kumar Patel, and many others for their constant moral support and camaraderie.

I would like to express respect and great admiration for my father, Mr. Ved Prakash Sharma, my mother Smt. Kamlesh Sharma, my wife Dr. Nitasha Bhardwaj, loving son Vedant and loving daughter Tushti for being the guiding light and encouraging force behind this endeavor. I would like to humbly dedicate this book series to my parents, Nitasha, Vedant, and Tushti.

I thank all the souls who helped me in this herculean task. Finally, I bow my head to the **ALMIGHTY** for all the blessings he has showered on me.

The Editors

Acknowledgments

We express our gratitude to CRC Press (Taylor & Francis Group) and the editorial team for their suggestions and support during completion of this book. We are grateful to all the contributors and reviewers for their illuminating views on each book chapter presented in *Additive Manufacturing in Industry 4.0: Methods, Techniques, Modeling, and Nano Aspects.*

Editors

Dr. Vipin Kumar Sharma received his doctoral degree with specialization in Nontraditional Machining Methods from the Thapar Institute of Engineering and Technology, Patiala with NIRF Ranking 23. He is currently working as Associate Professor in the Mechanical Engineering Department at IIMT University Meerut Uttar Pradesh, India. He has more than ten years of teaching and research experience related to mechanical engineering. Dr. Sharma has published 12 research papers in reputed journals with SCI indexing and ten in international and national conferences. Dr. Sharma also visited China for the presentation of a paper in the METADOR conference in 2019. His research interest includes hybrid composites with rare earth metals, casting methods, optimization techniques, and design of experiments. Dr. Sharma is also a reviewer of many reputed journals, such as Taylor & Francis publications (USA), Springer publications (Germany), SciEP publications (USA), *Journal of Materials Research Express* (IOP, Publishers, UK), and Advances in Materials Science (Hindawi). Currently, he is working in the fatigue analysis of hybrid composites reinforced with rare earth metals.

Dr. Ashwani Kumar received his PhD (Mechanical Engineering) in the area of Mechanical Vibration and Design. He is currently working as senior lecturer of Mechanical Engineering (Gazetted Officer Class II) at the Technical Education Department, Uttar Pradesh (under Government of Uttar Pradesh), India. He has worked as Assistant Professor in Department of Mechanical Engineering, Graphic Era University Dehradun India from July 2010 to November 2013. He has more than 12 years of research and academic experience in mechanical engineering. He is Series Editor of the book series *Advances in Manufacturing, Design and Computational Intelligence Techniques* and *Renewable and Sustainable Energy Developments* published by CRC Press, Taylor & Francis, USA. He is

Editor-in-Chief for *International Journal of Materials, Manufacturing and Sustainable Technologies* (*IJMMST*) and *Associate Editor for the International Journal of Mathematical, Engineering* and *Management Sciences* (*IJMEMS*) indexed in ESCI/Scopus and DOAJ. He is an editorial board member of four international journals and acts as a review board member of 20 prestigious (indexed in SCI/SCIE/Scopus) international journals with high impact factor i.e. *Applied Acoustics, Measurement, JESTEC, AJSE, SV-JME,* and *LAJSS*. In addition, he has published -96 research articles in journals, book chapters, and conferences. He has authored/coauthored and edited 18 books on mechanical and materials engineering. He has published two patents. He is associated with international conferences as invited speaker/advisory board/review board member. He has been recognized as *Best Teacher* and awarded for excellence in academic and research. He has successfully guided 12 BTech, MTech and PhD theses. In administration, he is working as coordinator for AICTE, E.O.A., nodal officer for the PMKVY-TI Scheme (Government of India), and internal coordinator for CDTP scheme (Government of Uttar Pradesh). He is currently involved in the research areas of machine learning, advanced materials, machining and manufacturing techniques, biodegradable composites, heavy vehicle dynamics, and coriolis mass flow sensor.

Dr. Manoj Gupta was a former Head of the Materials Division of the Mechanical Engineering Department and Director Designate of the Materials Science and Engineering Initiative at NUS, Singapore. He took his PhD from University of California, Irvine, USA (1992), and did postdoctoral research at the University of Alberta, Canada (1992). He is currently among the top 0.6% researchers as per the Stanford List, among the Top 1% of Scientists of the World Position by the Universal Scientific Education and Research Network, and among 2.5% among scientists as per ResearchGate. To his credit are the (1) Disintegrated Melt Deposition technique and (2) Hybrid Microwave Sintering technique, an energy-efficient solid-state processing method to synthesize alloys/micro-/nanocomposites. He has published over 600 peer-reviewed journal papers and owns two U.S. patents and one Trade Secret. His current h-index is 72, his RG index is 48.3, and his citations are greater than 18,000. He has also coauthored eight books, published by John Wiley, Springer, and MRF–USA. He is editor-in-chief/editor of 12 international peer-reviewed journals. A multiple award winner, he actively collaborates in/visits Japan, France, Saudi Arabia, Qatar, China, the United States, and India as a visiting researcher, professor, and chair professor.

Dr. Vinod Kumar received his Bachelor's Degree and Master's Degree, in Mechanical Engineering from Punjab University in 1995 and in 1999. He then received a Doctoral Degree specialization in Nontraditional Machining Methods from Thapar University. He is currently working as Associate Professor in the Mechanical Engineering Department at Thapar University. He joined the Thapar Institute of Engineering & Technology in June 2000 as Lecturer. His forte is practical know-how in nontraditional machining methods. Dr. Vinod Kumar has successfully completed an agencies-funded research project. Dr. Kumar has published 58 research papers in reputed journals and 32 in international and national conferences. His research interest includes powder metallurgy, casting methods, optimization techniques, design of experiments, among others. He has supervised five PhD theses and is currently supervising two PhD candidates in the area of powder metallurgy, casting methods, optimization techniques, and design of experiments. More than 50 Master's Theses have been completed so far under his supervision. Having a practical orientation and research inclination, he was the architect for Thapar University's Machining Labs. Dr. Kumar, as a captivating orator, has a long list of expert lectures delivered at renowned Institutes. Dr. Kumar is the reviewer of many reputable journals, such as Taylor & Francis publications, USA, Springer publications, Germany, SciEP Publications, USA.

Dr. Dinesh Kumar Sharma is working as Dean of Academics in IIMT University Meerut. He has worked as Deputy Director in the National Ecology and Environment Foundation (NEEF), Mumbai. Prior to his present position, he has also worked as Dean of Academics and HOD at different institutions/universities. He has completed his PhD (in Solar PV System, 2015), ME (Power Electronics, 2006), and BE (Electronics and Communication Engineering, 2003). He has more than 15 years of experience in teaching and research. Prof. Sharma has presented and published more than 22 papers in international journals and conferences of repute, e.g. IEEE, Springer, among others. He has authored a book on *Non-conventional Energy Resources* (2006). He is the recipient of Young Investigator Award and was invited to attend the 6th World Conference on Photovoltaic Energy Conversion (WCPEC-6) held in Kyoto, Japan, in 2014. He has also presented at the 13th Agricultural Science Congress (13ASC) at the University of Agriculture Sciences, Bangalore, in 2017. He is the member of various professional societies including ISEIS Canada, IAEng, Hong Kong, and ICTP Italy. Dr. Sharma is a reviewer and editorial board member of many international journals of repute such as IEEE, Taylor & Francis, among others. His areas of interest include solar energy, green buildings, and sustainable development.

Dr. Subodh Kumar Sharma received his BE, MTech, and doctorate in philosophy (PhD) in Mechanical Engineering from the National Institute of Technology Kurukshetra, Haryana. He has 21 years of research experience in academia. Presently he is working as a Professor in the Department of Mechanical Engineering, KIET Group of Institute, Delhi-NCR, Ghaziabad, India. He is also working as a Head of Industrial Research and Consultancy Centre in his parent institution. His research interests include thermal analysis and mechanical analysis of the internal combustion diesel engine, combustion chamber components, nanoadditives, and thermal barrier coatings. He is an author of two books: *Strength of Materials* and *Basic Fundamentals of Mechanical and Civil Engineering*. He has published many research papers in referred national and international journals and at conferences.

Contributors

Shalom Akhai
Department of Mechanical Engineering, Chandigarh Engineering College, Jhanjeri, Mohali, India

Kunal Arora
Department of Mechanical Engineering, Thapar Institute of Engineering and Technology, Patiala, Punjab, India

Ashish Bansal
Government Polytechnic College, Dewas, Madhya Pradesh, India

Harsh Kumar Bhardwaj
Department of Mechanical Engineering, Motilal Nehru National Institute of Technology Allahabad, Prayagraj, Uttar Pradesh, India

Shekhar Bhardwaj
KIET Group of Institutions, Ghaziabad, Uttar Pradesh, India

Mohd. Faizan
Department of Mechanical Engineering, Buddha Institute of Technology Gorakhpur, GIDA, Gorakhpur, India

Yatika Gori
Department of Mechanical Engineering, Graphic Era University Dehradun, Uttarakhand, India

Ajay Kumar Jena
Subharti Institute of Technology and Engineering, SVSU, Meerut, India

Ravinder Singh Joshi
Department of Mechanical Engineering, Thapar Institute of Engineering and Technology (deemed to be University), Patiala-147004 (Punjab), India

Priyanka Kaushik
Poornima Institute of Engineering and Technology, Jaipur (Rajasthan), India

Anish Kumar
Department of Mechanical Engineering, Maharishi Markendeshwar (deemed to be University Mullana- Ambala- 133207) (Haryana), India

Ashwani Kumar
Department of Mechanical Engineering, Technical Education Department Uttar Pradesh (Under Government of Uttar Pradesh) Kanpur, India

Jatinder Kumar
Department of Mechanical Engineering, National Institute of Technology, Kurukshetra-136119 (Haryana), India

Pardeep Kumar
Department of Mechanical Engineering, Chandigarh Engineering College, Jhanjeri, Mohali, India

Sachin Kumar
KIET, Ghaziabad, Uttar Pradesh, India

Sachin Kumar
Subharti Institute of Technology and Engineering, SVSU, Meerut, India

Vinod Kumar
Department of Mechanical Engineering, Thapar Institute of Engineering and Technology (deemed to be University), Patiala-147004 (Punjab), India

Arun Kumar Kushwaha
Department of Mechanical Engineering, Meerut Institute of Engineering and Technology, Meerut, India

Chandan Swaroop Meena
CSIR–Central Building Research Institute, Roorkee 247667, Uttarakhand, India

K. V. Ojha
KIET Group of Institutions, Ghaziabad, UP, India

Sunil Kumar Paswan
Department of Mechanical Engineering, Baba Farid College of Engineering and Technology, Bathinda 151001, India

Dinesh Kumar Patel
Department of Mechanical Engineering, IIMT Engineering College Meerut, Uttar Pradesh, India

Ram Bilas Prasad
Department of Mechanical Engineering, Madan Mohan Malaviya University of Technology, Gorakhpur, India

Deepak Sharma
Department of Mechanical Engineering, Thapar Institute of Engineering and Technology, Patiala, Punjab, India

Renu Sharma
Department of Physics, Maharishi Markendeshwar (deemed to be University Mullana-Ambala-133207) (Haryana), India

Sanjay Sharma
AKGEC Ghaziabad, Uttar Pradesh, India

Subodh Kumar Sharma
Department of Mechanical Engineering, KIET Group of Institute, Delhi-NCR, Ghaziabad, India

Vipin Kumar Sharma
Department of Mechanical Engineering, IIMT University Meerut Uttar Pradesh, India

Manpreet Singh
Department of Mechanical Engineering, Baba Farid College of Engineering and Technology, Bathinda 151001, India

Varun Pratap Singh
Department of Mechanical Engineering, University of Petroleum and Energy Studies, Dehradun, Uttarakhand, India

Darshan Srivastav
Department of Mechanical Engineering, Buddha Institute of Technology Gorakhpur, GIDA, Gorakhpur, India

Anand Tyagi
AKGEC Ghaziabad, Uttar Pradesh, India

Mansingh Yadav
Indian Institute of Technology Bombay, Maharashtra 400076, India

Chapter 1

Introduction to the flexible manufacturing system in Industry 4.0

Darshan Srivastav, Arun Kumar Kushwaha,
Mohd. Faizan, and Vipin Kumar Sharma

DOI: 10.1201/9781003360001-1

1.1 INTRODUCTION

A *manufacturing system* is defined as a system consisting of a collection of integrated equipment, which may be machine tools, material handling and work positioning devices, and a computer, and of a human resource to keep the equipment running. The integrated equipment performs the necessary processing and assembly operations required to transform an initial raw material into a finished part [1].

The capabilities of traditional manufacturing systems are not at par with modern manufacturing requirements like customized products, larger volumes and variety, sustainability, customer responsiveness, excellent quality and flexibility, and automation in processes. To accomplish these challenges, various manufacturing philosophies like *lean manufacturing*, *agile manufacturing*, *sustainable manufacturing*, and *automated manufacturing systems* have evolved over time [2].

The concept of lean manufacturing is motivated by the idea of waste elimination; agile manufacturing incorporates the capability of greater customer responsiveness; and sustainable manufacturing is the concept that enables the industry to carry out operations in an environment-friendly way with the objective of judicious use of resources, keeping in mind the needs of future generations. While these three manufacturing philosophies deal with process management, customer orientation, and sustainability, automated manufacturing incorporates the technological transformations made in manufacturing systems so as to achieve the goals of mass production, quality goals, flexibility, and lessened human involvement.

The capability of FMS to produce parts in different varieties and different mixes of part styles at a time, along with its responsiveness to the changing demand trends, gives it the title of *flexible* [1].

1.2 FLEXIBLE MANUFACTURING SYSTEMS

The flexible manufacturing system is a system that consists of machine cells that are based on the principle of group technology (GT). The FMS is comprised of an integration of several machines that are controlled by computing devices, a system for material handling with high level of automation, and a system for the storage and recovery of parts. This model of a manufacturing system is based on the philosophy of the effective control of material flow through a network of versatile processing stations using an efficient and adaptable material handling and storage system. The machine cells are arranged along an automatic material handling system, which can be conveyors, trucks, automatic guided vehicles, or the like. FMS is flexible by virtue of its capability of processing a variety of different part styles simultaneously at the various workstations, and the mix of part styles, and volumes of production can be adjusted in response to changing demand patterns. The FMS is one of the most used technologies in the manufacturing of medium-sized volumes of products in moderate variety.

1.2.1 Flexibility in manufacturing

Flexibility refers to the degree to which a manufacturing system is capable of dealing with the variations in the varieties and the volumes of the products that it produces. Flexibility not only means quick response to ever changing customer demands, but also it indicates the ability of the system to undergo the necessary transformations and adjust to changing conditions. Flexibility in a manufacturing system is classified under the following eight types [3]:

1. *Machine flexibility:* The ease of undergoing the changes required to produce a part family. It may include the time to repair or replace the worn-out tools, tool changing time in a tool magazine, the time to set up newer fixtures, among meeting other needs.
2. *Part flexibility:* The ability to quickly and economically adjust the requisite changes in order to produce a new set of parts. This flexibility reflects the company's potential responsiveness to competitive market fluctuations. It can be attained by means of efficient and automated production planning and control.
3. *Process flexibility:* The ability to produce parts individually instead of in batches. Each part in a part family can be produced in different ways by changing the material used or the operations performed. A subset of this flexibility is *Operations Flexibility*.
4. *Routing flexibility:* The ability to produce parts by processing them through several paths or routes. This improves the ability to handle breakdowns. It is possible when the same operation can be performed on more than one machine. Such flexibility makes a manufacturing system robust. It can be attained by allowing automatic rerouting of products, which is made possible by grouping machines into machine cells.
5. *Volume flexibility:* It refers to the ability of a manufacturing system to perform profitably even at different production volumes. It enables the system to cope with the repercussions of issues like changes in market demand, catastrophic fluctuations, or recession. A high level of automation is responsible for such capability.
6. *Operation flexibility:* The ability to change the existing order of operations in the processing of a part type. It requires real-time decision making that depends on the current state of the system (e.g. which machine station is idle, empty locations for storage, etc.).
7. *Expansion flexibility:* The capability of easily and modularly expanding the manufacturing system as and when needed. This flexibility can be

attained by having a non-process-oriented layout, a flexible automated material handling and storage system, and modular machine cells with pallet changers.

8. *Production flexibility:* The ability of a manufacturing system to produce different part types in different volumes and part mixes. It is a measure of the level of automation and technology existing in the system. This flexibility can only be attained by incorporating all the preceding flexibilities.

Ideally, a flexible manufacturing system should possess all these flexibilities. Not all these flexibilities are individual, but one or more flexibilities are often interdependent. Figure 1.1 represents this interdependence. An intersection between two circles indicates that the two flexibilities depend on each other.

1.2.1.1 Components of FMS

A flexible manufacturing system typically consists of the following elements:

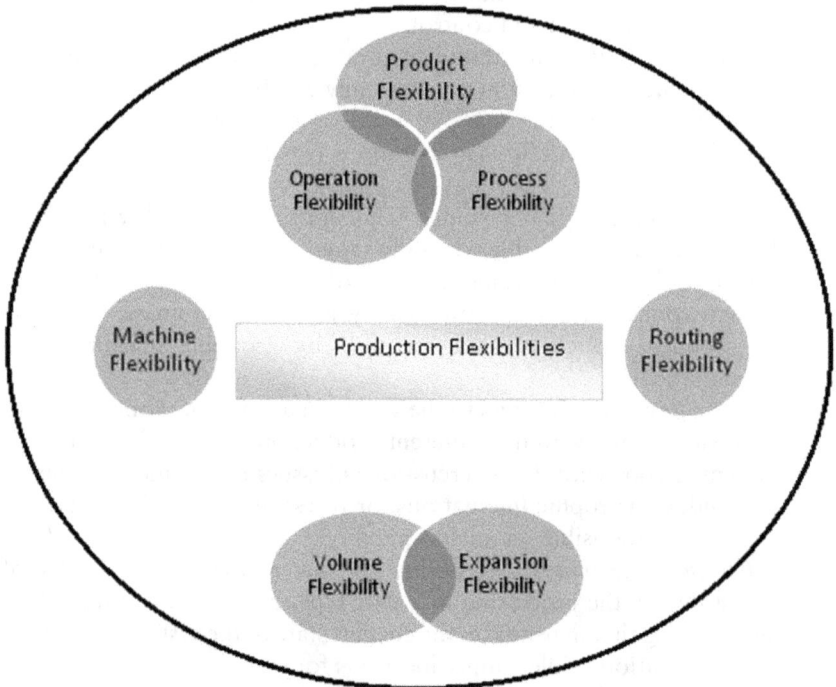

Figure 1.1 Relationship between different types of flexibilities in manufacturing.

- Numerical control (NC) machine tools
- Automated material handling systems (AMHS)
  - Conveyor belts
  - Automated guided vehicles
  - Automated storage and retrieval system (AS/RS)
- Industrial robots
- Control system with computers

1.2.2 Material handling and logistics in FMS

In general, *logistics* refers to the management of the flow of things from the source of origin to the destination of consumption to meet the customer requirements. It takes care of the processes of the acquisition, movement, storage, and distribution of materials to satisfy customers.

- *External logistics:* The transportation and other related activities that occur between different geographical locations outside a factory. Traditional modes of transportation like railways, trucks, aircraft, ships, and pipelines facilitate this movement
- *Internal logistics:* The material handling and movement within a facility. Modern technologies that facilitate this movement are trucks, cranes, hoists, conveyors, rail-guided vehicles (RGVs), and automated guided vehicles (AGVs).

Material handling can be defined as the safe and secure transport, storage, and control of the flow of materials and products throughout the entire process of production and distribution, consumption, and disposal. The Material Handling Industry of America [41].

It is estimated that 20–25% of the total manufacturing is involve the material handling process and equipment. However, this proportion varies depending on the type of manufacturing process and the level of automation involved. Hence, it is necessary to ensure that material handling is performed efficiently, cost-effectively, accurately, safely, and in a timely manner.

1.2.2.1 Classification of material handling equipment

The equipment can be classified into five major types:

1. *Transport equipment:* To move materials within a facility, warehouse, or factory. Such equipment includes trucks, conveyors, cranes and hoists, RGVs, and AGVs.
2. *Positioning equipment:* To handle parts at a particular location. Positioning is accomplished by using industrial robots, hoists, etc.

Table 1.1 Selection of material handling equipment on the basis of type of plant layout

Plant layout	Material handling equipment
Fixed-position layout	Cranes, hoists, industrial trucks
Process layout	Hand trucks, forklifts, AGVs, RGVs
Product layout	Conveyor for product flow
	Trucks to deliver parts to stations

3. *Unit load formation equipment:* This includes material-holding containers and loading and packaging equipment. Examples are wooden pallets, pallet box, tote pans, etc.
4. *Storage equipment:* To store and hold materials and provide access to them as and when required. It is further classified into two sub-categories: (a) conventional storage and (b) automated storage. Conventional storage serves the purpose of bulk storage by using racks, shelves and bins, drawers, etc. Conventional storage methods are labor intensive. Automated storage systems, like an *automated storage-retrieval system*, are capable of reducing this manual labor to a large extent.
5. *Identification and control equipment:* The task of identifying and tracking the material being stored or moved. Technologies like bar codes, quality response codes, and *radio-frequency identification* (RFID) are usually employed for the purpose.

The type of equipment used has to be selected depending on the type of plant layout.

Table 1.1 represents material handling equipment selection on the basis of type of plant layout.

1.3 AUTOMATED GUIDED VEHICLES

An automated guided vehicle (AGV) is a self-propelled, computer-controlled, unmanned vehicle, usually powered by battery; which is guided along pre-defined pathways [4]. The AGV is an automatic guided cart that follows a predefined path. This mode of material handling is being largely used in industries and other sites of distribution. In general, AGV is a transport system that can operate driverless and is used for the movement of materials, the movement being horizontal. It can be used for both internal and external transport. The AGVs are capable of navigating their predetermined paths by means of state-of-the-art sensor technology consisting of vision cameras, magnets, and laser-based sensors.

The automated guided vehicles are classified into three different categories:

1. *Towing vehicles or driverless trains:* Used for moving heavy loads over large distances
2. *Pallet trucks:* For moving palletized loads over predetermined paths
3. *Unit load carriers:* Used for transporting unit loads between stations in a facility

Scheduling forms a vital component in the planning and control of flexible manufacturing systems (FMS). Scheduling is the process of assigning and allocating a fixed volume of resources that may be the plant and machinery resources; for different jobs and operations while manufacturing several products at the same time. Scheduling is one of the most important tasks that must be considered in the design of an efficient FMS setup. The following section presents a literature survey of significant research works carried out in this area.

1.4 STATE-OF-THE-ART LITERATURE REVIEW

Different trends and paradigms in modern manufacturing in the present scenario are well presented by Nambiar [2]. A comparison between lean, agile, and automated manufacturing is done. Flexible manufacturing systems (FMSs) belong to the class of automated manufacturing, with a compromise between a higher level of automation and flexibility at the same time [1]. The need for and structure of efficient production planning in such systems is well presented in the book by Bedworth and Bailey [5]. Different types of automated manufacturing systems, FMS classification, layouts, and types of flexibility have been presented in greater details in Groover [1] and Sethi, Stecke, and Browne [3]. Stecke and Soldberg [6] analyzed the important concerns and different problems encountered in order to establish a full-fledged FMS.

In upcoming subsections, significant works and contributions in sequencing–scheduling in production control, AGVs and vehicle routing problems, and application of metaheuristics in scheduling and routing problems – have been presented in tabular form.

1.4.1 Review of sequencing – scheduling and design parameters of FMS

Tetzlaff [12] compiled different aspects of design considerations for planning, tooling, and modeling in flexible manufacturing systems in greater detail. Different programming and heuristic-based tools are proposed and implemented as examples. Relevant research work is tabulated as in Table 1.2.

Table 1.2 Survey of sequencing, scheduling, design, and modeling of FMS

S. No.	Author(s)	Year	Research findings and contribution
1	Stecke and Soldberge [6]	1982	Different concerns in establishing a full-fledged FMS. A constraint-based nonlinear programming method is used for machine grouping and loading problems.
2	Sethi, Browne, and Stecke [3]	1984	Different grounds for classification of FMS on the basis of different types of flexibilities
3	Bedworth and Bailey [5]	1987	Descriptions of production control, production planning, scheduling, material requirement planning, etc. are presented in detail. Computer programming approach is proposed to solve problems related to these areas.
4	Adams, Balas, and Zawack [14]	1988	A heuristic-based approach known as shifting bottleneck problem is proposed, which is found to be very efficient in solving job shop problems.
5	Tetzlaff [12]	1990	Design considerations in planning, tooling, and modeling of flexible manufacturing systems are discussed. Different programming and heuristic based tools are proposed.
6	Hansmann and Hoek [15]	1997	A neighborhood search technique is used for a job shop scheduling problem in FMS.
7	Ficko, Brezocnik and Balic [16]	2004	A Genetic Algorithm (GA) model is proposed for FMS. Facility layout problem is addressed. Components like number of trips, transportation costs, and distance between devices are used for problem formulation.
8	Geiger, Uzsoy, and Aytuğ [17]	2006	Different priority dispatching rules (PDRs) in scheduling are discussed, and a rule discovery system is proposed. These rules are implemented on single machine problem.
9	Blazewicz et al. [18]	2007	Design considerations in scheduling of FMS. Heuristic approaches, like the binary search technique for both static and dynamic scheduling, are presented. Combined tasks like scheduling and routing, job and vehicle scheduling are also discussed in detail with algorithms.
10	Baker and Trietsch [13]	2009	Theoretical as well as practical aspects of sequencing and scheduling in manufacturing are discussed. Different heuristics approaches have been formulated for the scheduling problems.

1.4.2 Review of AGV and vehicle routing problems in production

Kusiak and Cyrus [19] considered various scheduling problems associated with AGVs. Some of these problems are determining the shortest path for travel, minimum number of AGVs needed, avoidance of traffic congestion and conflicts, etc. Some heuristics-based algorithms for addressing these problems are also presented in this paper. Fazlollahtabar and Saidi [11] have detailed different AI-based models for the modeling and simulation of AGV for scheduling and routing problems. A review of vehicle routing problems and related research work is presented in Table 1.3.

Table 1.3 Survey of AGV and routing problems in literature

S. No.	Author(s)	Year	Research findings and contribution
1	Kusiak and Cyrus [19]	1985	Various routing and scheduling problems associated with AGVs are discussed. Heuristic-based algorithms for addressing these problems are presented to solve these problems with the objective of finding the optimal number of AGVs needed in order to optimize the network design.
2	Mahadevan and Narendran [20]	1990	Design considerations of issues in AGV-based material handling system are presented in detail. Scheduling techniques are discussed with the objective of traffic problems.
3	Krishnamurthy, Batta, and Karwan [21]	1991	The static routing problem in AGV is addressed. The assignments of AGVs being predetermined, the objective is to minimize the makespan while avoiding conflict and congestion on the path. AGVs are bidirectional, and column generation heuristics are modeled.
4	Li, Adriaansen, Udding, and Pogromsky [22]	2011	A case study of AGVs in a container transport environment is considered. An efficient model considering the number of layouts is developed, and an algorithm is formulated for the routing of AGVs considering the objective of minimum distance of travel, minimum transport time, and traffic control.
5	Adulyasak, Francois, and James [23]	2014	A comprehensive review of vehicle routing in inventory management and lot sizing problem is presented. Branch and Bound technique and several heuristic based approaches are used.
6	Fazlollahtabar and Saidi [11]	2015	Different AI-based models for modeling and simulation of AGV for scheduling and routing problems are discussed. Several mathematical models like stochastic model, simulation and reliability model, etc. are formulated for the AGV scheduling.

1.4.3 Review of metaheuristics and their application in scheduling and routing problems

An excellent explanation of the difficulties or hardness that are generally encountered in complex optimization problems is presented by Chopard and Tomasini [25]. The idea of P and NP complete problems is also discussed. Different metaheuristics and their development are also discussed. Mirjalili, Dong, and Lewis [24] have presented different cases of complex optimization problems, as well as applied metaheuristics for their solution. The level of computational complexity associated with a traveling salesman problem is described in detail. Table 1.4 represents a review of research works on metaheuristics based approach for sequencing and scheduling problems.

Table 1.4 Application of metaheuristics in scheduling and routing problems

S. No.	Author(s)	Year	Research findings and contribution
1	Kirkpatrick, Gelatt, and Vecchi [26]	1983	An analogy between multiple degrees of freedom systems in thermal equilibrium and the combinatorial optimization problems is developed. A new optimization tool known as simulated annealing inspired by this analogy is proposed.
2	Kennedy and Eberhart [27]	1995	A new optimization technique inspired by the swarm behavior of birds flocking, fish schooling, etc. is developed. The particle swarm optimization is formulated for nonlinear continuous functions.
3	Mitchell [28]	1996	History and development of the Genetic Algorithm is discussed in detail. The operators of GA and their application on different classes of optimization problems is presented.
4	Dorigo and Gamberdella [29]	1996	Ant-system-based on the ant colony technique is used to address the traveling salesman problem so as to find the best tour.
5	Shi and Eberhart [30]	1998	A new parameter known as inertia parameter is introduced. It is found that, with higher inertia weight, the PO can explore a larger search area.
6	Dorigo and Caro [32]	1999	A new algorithm advancing from the idea of Ant systems based on the foraging behavior of ants known as Ant Colony Optimization is formulated, and examples of its application are presented.
7	Dorigo, Bonabeau, and Theraulaz [33]	2000	The phenomenon of stigmergy in organisms is discussed. Models are developed for this study of real ants and its significance is analyzed.
8	Kumar, Tiwari, and Shankar [34]	2003	Extent of complexity in scheduling problem in FMS is analyzed and ant colony optimization technique is proposed.

Table 1.4 Cont.

S. No.	Author(s)	Year	Research findings and contribution
9	Sankar, Ponabalan and Rajendran [35]	2003	A FMS with 5 cells and 2 AGVs is considered. A 9-machine and 43-part scheduling problem is defined for which a multiobjective Genetic Algorithm is used.
10	Jerald, Asokan, Prabaharan, and Sarvanan [36]	2004	Nontraditional optimization techniques like the Genetic Algorithm, Memetic Algorithm, Simulated Annealing are compared, and Particle Swarm Algorithm is attempted on the 9-machine, 43-part scheduling problem in FMS.
11	Babu, Jerald, Haq, Luxmi, and Vigneswaralu [37]	2009	Simultaneous scheduling of machines and AGVs in FMS environment. Two AGVs are considered. Scheduling of AGVs is described to be an integral part of FMS scheduling. Differential evolution algorithm is chosen for the purpose.
12	Burnwal and Deb [38]	2012	Cuckoo Search technique is used for scheduling problem in the 9-machine, 43-part FMS layout.
13	Haq, Karthikeyan, and Dinesh [39]	2003	A problem case of integrated production schedule and AGV schedule in a FMS layout with 2 parts, 9 machines, and an AGV is considered. GA and SA (Simulated Annealing) techniques are implemented, and results are compared.

1.5 FINDINGS FROM LITERATURE

- Scheduling of AGVs should be considered an integral part of the FMS scheduling problem. An approach for integrated job shop scheduling and AGV routing is a good way to ensure optimal production planning in a FMS environment.
- Both job shop scheduling as well as routing considerations in a FMS form a problem of an NP complete class because of the computational complexity involved.
- Modern heuristics- and computational-intelligence-based approaches have to be implanted to bring out some more efficient solutions for this combinatorial problem of scheduling and vehicle routing.
- Metaheuristics are often the only tools that researchers resort to, in solving optimization problems with a computational complexity of the order of nondeterministic polynomial time.

Considering these four implications from the literature, an integrated scheduling and AGV routing problem is presented in next section. The combinatorial problem instance is derived from the literature [13] and is addressed with the Ant Colony Optimization technique.

Table 1.5 FMS loop layout configuration

Layout type	No. of machines	No. of parts	Load/Unload stations	No. of AGVs
U-shaped	9	2	1 each	1

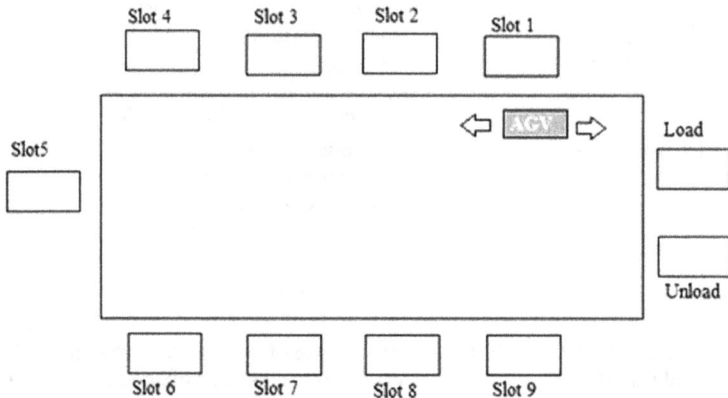

Figure 1.2 Layout of 2 parts, 9 machine FMS with AGV.

1.6 MATERIALS AND METHODS

1.6.1 Problem definition and methodology

Loop layout is one of the most common layouts in a flexible manufacturing system. The AGV operates in a bidirectional manner. A loop layout design can be represented as permutation of machines say $(M_1, M_2, M_3 \ldots M_n)$ with a prefix of loading and unloading stations. For every part, a well-defined part route is specified. A machine part incidence table shows the incidence of parts on machine stations. Usually, positions around the loop where the machine stations are situated are identified as *slots*.

A configuration of a FMS with 2 parts, which are to be processed on 9 machines, is presented in the literature [13]. The FMS consists of a loading station for holding the raw parts and an unloading station for holding the finished part. These parts are moved from these stations to the machine stations with the help of an automated guided vehicle that can move in a bidirectional manner. The configuration details are presented in Table 1.5, and the layout is shown in Figure 1.2. The machine-part incidence is shown in Table 1.6, which represents the incidence or arrival of a particular part on different machines in the order defined by its processing requirement.

Table 1.6 Machine part incidence matrix

Part type	Operating Sequence								
	1	*2*	*3*	*4*	*5*	*6*	*7*	*8*	*9*
P₁	M_2	M_5	M_3	M_1	M_8	M_4	M_6	M_9	M_7
	M_2	M_6	M_3	M_5	M_1	M_4	M_9	M_7	M_8
P₂	M_6	M_2	M_1	M_7	M_8	M_3	M_4	M_9	M_5
	M_6	M_9	M_5	M_3	M_1	M_4	M_7	M_8	M_2

Table 1.7 Processing times of the parts on the machines

Part type	Processing Time								
	M_1	M_2	M_3	M_4	M_5	M_6	M_7	M_8	M_9
P₁	62	9	89	87	10	66	95	49	82
P₂	30	95	58	31	58	60	18	8	5

The processing times of the parts on the machines is shown in the Table 1.7.

Given the processing requirements of jobs in terms of the part–machine incidence matrix, the best routing of the parts for ensuring routing flexibility is determined using the well-known priority dispatching rules. [17]

The optimum part route so obtained is provided as an input to schedule the material handling system. Machines are allocated to the slots in the layout in the best possible schedule so as to obtain an optimum facility layout. The machine allocation problem in the layout is addressed considering the following bicriteria objective;

1. *Shortest path of travel:* The total distance traveled by the AGV during the entire operation for processing all the jobs is minimum.
2. *Least number of backtrackings*

A combined objective function that includes both these objectives is represented by equation (1.1a):

Minimize $Z = w_1D + w_2B$ .... (1.1a)

where

$$D = \sum_{i=1}^{p}\sum_{j=1}^{M}\sum_{k=1}^{M} d_{ijk}; \; d_{ijk} =; \text{ if } j = k$$

$$B = \sum_{i=1}^{P}\sum_{j=1}^{M}\sum_{k=1}^{M} b_{ijk}; \quad b_{ij} = \begin{cases} 1, & if \ j > k \\ 0, otherwise \end{cases} \ldots \qquad (1.1b)$$

and d_{ijk} = distance traveled by the AGV in moving a part i from machine j to machine k.

b_{ijk} = number of backtrackings occurring during the move.

D = total distance traveled by the AGV due to the movements occurring during the entire operation of processing all the P parts.

B = total number of bactrackings occurred due to the movement of the AGV during the entire operation of processing all the P parts.

w_1 and w_2 = normalized weights assigned to each parameter representing the objective.

1.6.2 Objectives

- To schedule the FMS for the optimum processing sequence for each of the parts to be processed in order to minimize makespan.
- To determine the optimum route for the AGV considering the criterion of the shortest path of travel and minimum number of backtrackings
- To optimize the facility allocation for a given facility layout in a FMS environment

1.6.3 Methodology

Figure 1.3 describes the methodology to address the combinatorial optimization problem.

One of the objectives of this research is to determine the optimum processing sequence of the parts so as to minimize the makespan in an FMS environment for a given part–machine incidence data. The well-known *priority dispatching rules* (PDRs) are implemented for this work.

Another, major objective of this work is to find an optimal route for the AGV depending on the obtained schedule for the parts and thereby optimize the facility layout so as to maximize factory utilization. This problem is addressed using a heuristic *Ant Colony Optimisation* (ACO) technique.

1.6.4 Part–machine scheduling using shortest processing time (SPT) rule

The following points must be considered while implementing the SPT and FIFO rules:

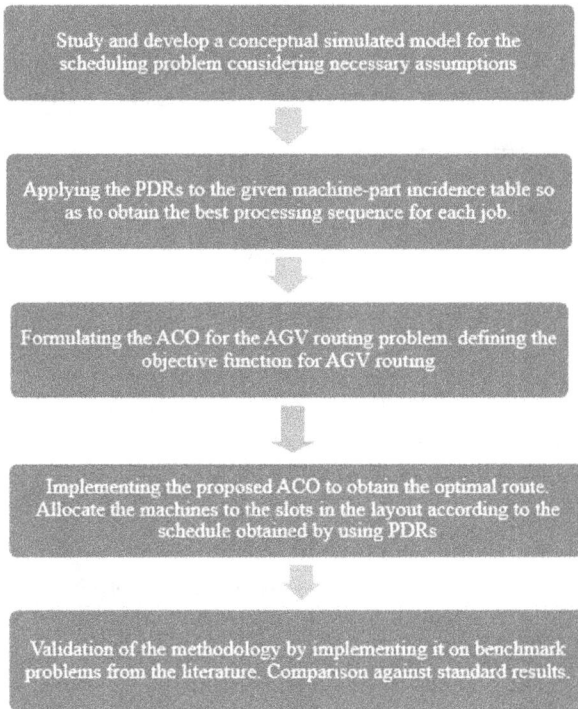

Figure 1.3 Methodology of the proposed research.

1. Every machine has to be assigned one operation of each of the parts.
2. Among two machines available for an operation, the one that takes less time to process is scheduled for that operation, provided the rule in point (1) is not violated.
3. The priorities are also based on the FIFO rule, provided the rule in point (1) is not violated.

Schedule for Part 1: The assignment of operations on machines by implementing the SPT + FIFO rules is obtained in the first trial as $M_2 \rightarrow M_5 \rightarrow M_3 \rightarrow M_1 \rightarrow M_8 \rightarrow M_4 \rightarrow M_6 \rightarrow M_9 \rightarrow M_7$. Operation 2 of part1 had two alternatives, M_5 and M_6. Processing time on M_5 is 10, which is less than the processing time on M_6, which is 66. Hence M_5 is selected. Also, the option for M_5 comes first for operation 2 and then for operation 4. The schedule obtained by considering these two rules, in the first trial itself, does not violate the first rule (1).

Schedule for Part 2: The problem case for part 2 can be summarized in Table 1.8.

Table 1.8 Scheduling problem for illustration 1

Operation no.	O1	O2	O3	O4	O5	O6	O7	O8	O9
Machine no.	M_6	M_2/M_9	M_1/M_5	M_7/M_3	M_8/M_1	M_3/M_4	M_4/M_7	M_9/M_8	M_5/M_2
Processing Time	60	95/5	30/58	17/58	8/30	58/81	61/17	5/8	58/95

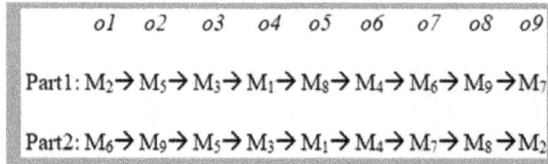

$$\begin{array}{ccccccccc} o1 & o2 & o3 & o4 & o5 & o6 & o7 & o8 & o9 \end{array}$$

Part1: $M_2 \rightarrow M_5 \rightarrow M_3 \rightarrow M_1 \rightarrow M_8 \rightarrow M_4 \rightarrow M_6 \rightarrow M_9 \rightarrow M_7$

Part2: $M_6 \rightarrow M_9 \rightarrow M_5 \rightarrow M_3 \rightarrow M_1 \rightarrow M_4 \rightarrow M_7 \rightarrow M_8 \rightarrow M_2$

Figure 1.4 Final schedule for Part 1 and Part 2.

PDRs are implemented in three to four different trials:

Trial 1: $M_6 \rightarrow M_9 \rightarrow M_1 \rightarrow M_7 \rightarrow M_8 \rightarrow M_3 \rightarrow M_4 \rightarrow$ _____

Trial 2: $M_6 \rightarrow M_9 \rightarrow M_5 \rightarrow M_7 \rightarrow M_1 \rightarrow M_3 \rightarrow M_4 \rightarrow M_8 \rightarrow M_2$

Trial 3: $M_6 \rightarrow M_9 \rightarrow M_5 \rightarrow M_3 \rightarrow M_1 \rightarrow M_4 \rightarrow M_7 \rightarrow M_8 \rightarrow M_2$

In trial 1 for operation 8, no machine is available as both M_9 and M_8 have already been assigned. This violates the rule (1). To rectify this, the schedule for either operation 2 or operation 5 has to be modified. Since machine M_1 has already been assigned for operation 3, hence it cannot be assigned to operation 5. Hence, changes for operations 3, 2, and 5 have been attempted in successive trials. Different schedules have been obtained in successive trials, and the completion time of Part 2 is computed for each trial. The trial that gives minimum completion time is the best one. Clearly, both trial 2 and trial 3 are feasible and give a total completion time of 392. Hence, either of the two can be considered the optimal schedule for part 2.

Thus the final optimal schedule for both parts is shown in Figure 1.4.

1.6.5 Ant Colony Optimization for AGV routing problem

The objective of the problem can be explained simply as how to assign the 9 machines to the different slots available in the FMS layout, so that the length of the path traveled by the AGV while moving these parts to different machines according to the optimal schedule obtained in the previous section is minimum.

Table 1.9 Distance matrix for the FMS layout in illustration I

Slots	L	U	S1	S2	S3	S4	S5	S6	S7	S8	S9
L	0	6	4	6	8	10	16	16	14	12	10
U	6	0	10	12	12	14	16	16	8	6	4
S1	4	10	0	2	4	6	12	18	18	16	14
S2	6	12	2	0	2	4	10	16	18	18	16
S3	8	14	4	2	0	2	8	14	16	18	18
S4	10	16	6	4	2	0	6	12	14	16	18
S5	16	16	12	10	8	6	0	6	8	10	12
S6	16	10	18	16	14	12	6	0	2	4	6
S7	14	8	18	18	16	14	8	2	0	2	4
S8	12	6	16	18	18	16	10	4	2	0	2
S9	10	4	14	16	18	18	12	6	4	2	0

Usually, in vehicle routing problems, a general trend of starting with a *distance matrix* is found in the literature. The distance matrix is nothing but a representation of the distance between different stations in a layout or a facility. The distance matrix for the considered 2 parts–9 machines FMS layout is shown in Table 1.9. Clearly, this problem resembles the nature of a *traveling salesman problem* (TSP) with the salesman being analogous to the AGV and the cities analogous to the slots for machine stations. A graph can be constructed using these 11 points (9 slots plus 1 load and 1 unload station). Coordinates of these points can be computed using the distance matrix. In order to solve this *NP hard* problem, the Ant Colony Optimization technique is implemented. Pseudo codes are generated, and the results about the optimal tour for the ants are obtained.

The entire implementation and coding process is done using MATLAB software. MATLAB R2013a (64-bit win64) is installed on a laptop with an i3-based 2.30 GHz processor configuration.

The modeling of an optimization problem for ACO consists of three phases: construction phase, pheromone model, decision model. The construction phase involves the initialization of the problem with the ant system, generation of the ant colony and subcolonies, etc. In pheromone modeling, the mathematical formulation of the amount of pheromone concentration is done, which is based on the information of the distance and quality of the path. The selection of paths is made depending on the probability rules. The flow chart for the AGV routing problem using ACO is presented in Figure 1.5. The pseudo codes are given following the flow chart.

1.6.6 Pseudo codes of ACO for solving AGV routing problem

for number of iterations = max. no. of iterations.
  create colony.
  calculate the fitness value of each ant.
  find the best ant (queen).

  update pheromone.
  compute evaporation.
  visualize the best tour.
end

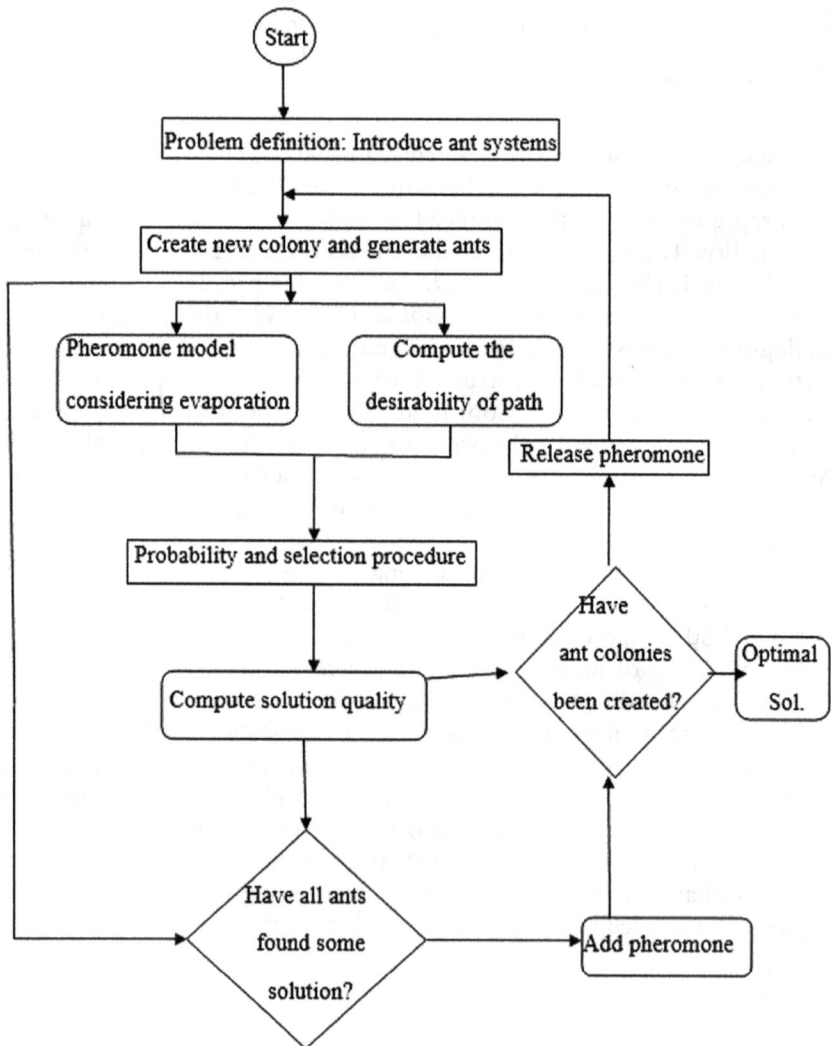

Figure 1.5 Flow chart for Ant Colony Optimization.

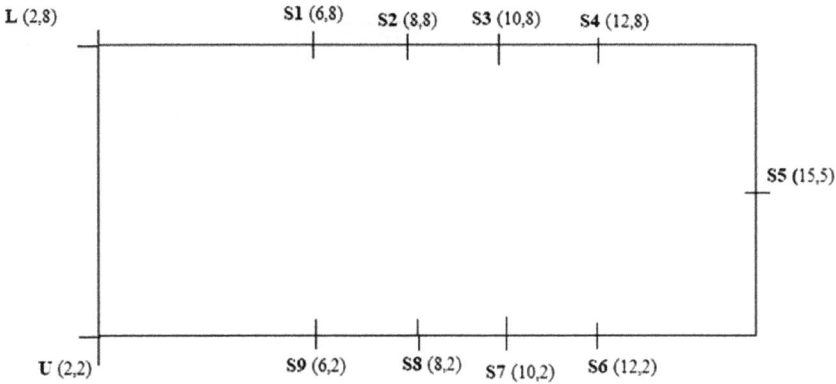

Figure 1.6 Coordinates of simulated FMS layout.

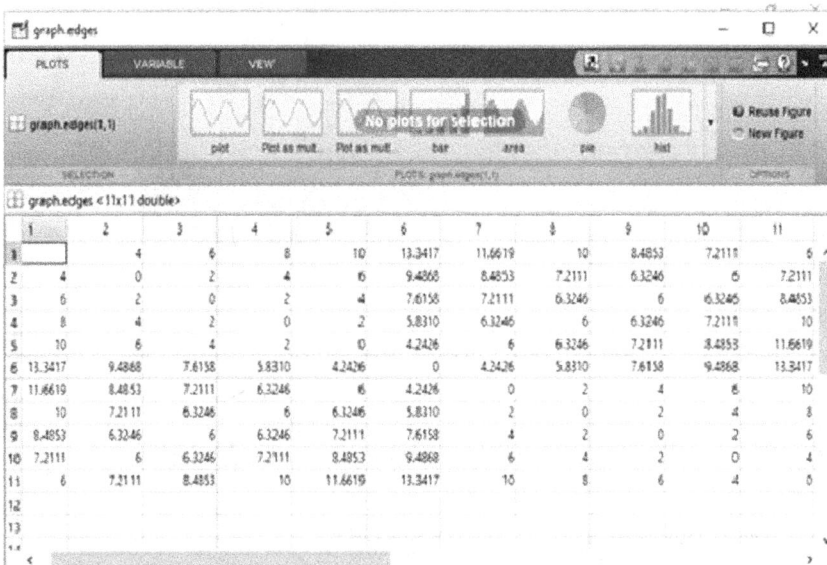

	1	2	3	4	5	6	7	8	9	10	11
1		4	6	8	10	13.3417	11.6619	10	8.4853	7.2111	6
2	4	0	2	4	6	9.4868	8.4853	7.2111	6.3246	6	7.2111
3	6	2	0	2	4	7.6158	7.2111	6.3246	6	6.3246	8.4853
4	8	4	2	0	2	5.8310	6.3246	6	6.3246	7.2111	10
5	10	6	4	2	0	4.2426	6	6.3246	7.2111	8.4853	11.6619
6	13.3417	9.4868	7.6158	5.8310	4.2426	0	4.2426	5.8310	7.6158	9.4868	13.3417
7	11.6619	8.4853	7.2111	6.3246	6	4.2426	0	2	4	6	10
8	10	7.2111	6.3246	6	6.3246	5.8310	2	0	2	4	8
9	8.4853	6.3246	6	6.3246	7.2111	7.6158	4	2	0	2	6
10	7.2111	6	6.3246	7.2111	8.4853	9.4868	6	4	2	0	4
11	6	7.2111	8.4853	10	11.6619	13.3417	10	8	6	4	0
12											
13											

Figure 1.7 Output of graph.edge.

1.6.7 Simulation of routing problem into ant system

On the basis of the distance matrix, different slots of the FMS layout can be assigned some coordinate to form a simulated layout model, as shown in Figure 1.6.

These coordinates are used to define the graph nodes of the ant colony. The edge matrix output is shown in Figure 1.7.

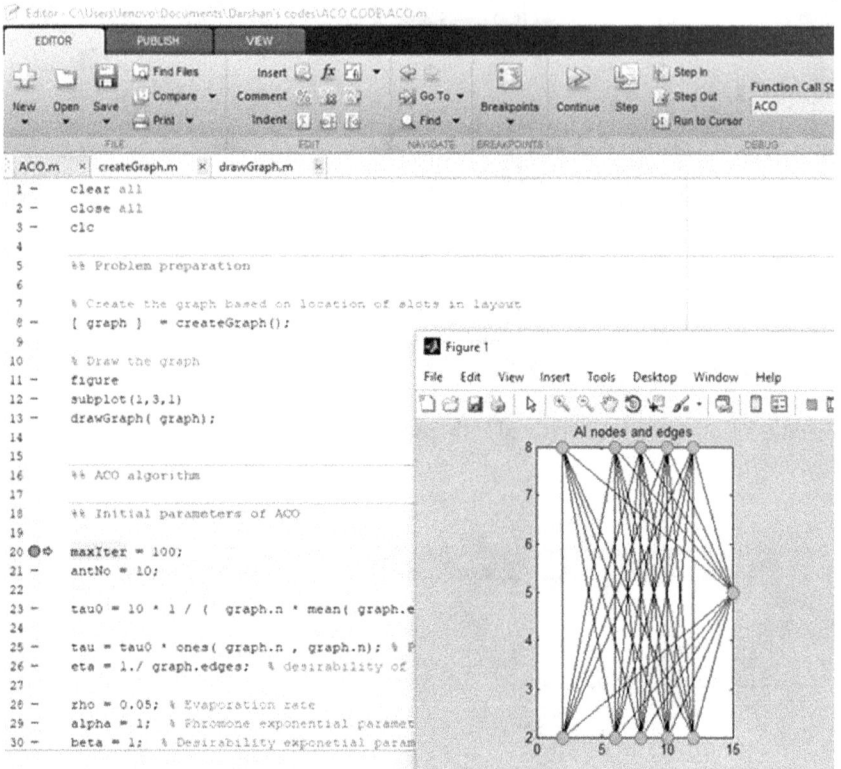

Figure 1.8 Output of drawGraph.m – all nodes and edges.

The function drawGraph draws the graph by connecting each node to all possible nodes. In turn, it also gives an idea of all possible tours that can be obtained in this problem. The graph drawn is shown in Figure 1.8.

1.7 RESULTS AND DISCUSSIONS

The best tour obtained is the one that is indicated by the maximum number of pheromones. The best tour, referred to as "the queen," is obtained gradually in a number of iterations, clearly 20 iterations with the fitness value of 34.4853, which is the shortest length of tour. The best tour so obtained indicates the following path:

$$L \to S3 \to S2 \to S9 \to S1 \to S6 \to S4 \to S5 \to S8 \to S7 \to U$$

If the layout is arranged in this order, then the machine assignments to the slots would look like the one shown in Figure 1.9.

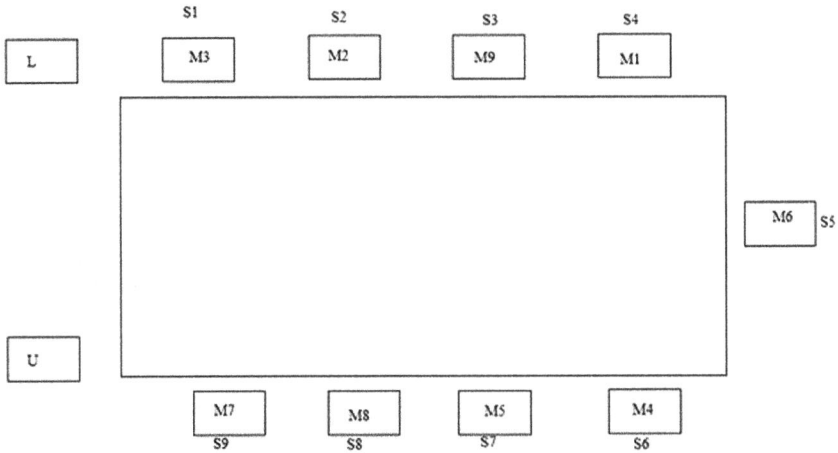

Figure 1.9 Optimal layout obtained by ACO.

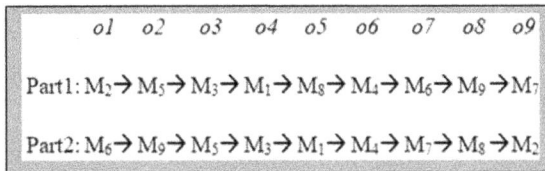

Figure 1.10 Final optimal schedule.

Considering the dual objective of AGV routing discussed in section 1.2.1; the objective function:

Minimize $Z = w_1D + w_2B$ .... $\qquad$ (1.1a)

where

$$D - \sum_{i=1}^{p}\sum_{j=1}^{M}\sum_{k=1}^{M}d_{ijk}; \; d_{ijk} = 0; \text{ if } j = k$$

$$B = \sum_{i=1}^{p}\sum_{j=1}^{M}\sum_{k=1}^{M}b_{ijk}; \; b_{ij} = \begin{cases} 1, & \text{if } j > k \\ 0, & \text{otherwise} \end{cases} \cdots \qquad (1.1b)$$

and

d_{ijk} = distance traveled by the AGV in moving a part i from machine j to machine k.

b_{ijk} = number of backtrackings occurring during the above move.

Table 1.10 Objective value for the best tour

	Part 1	Part 2
D	94	86
B	2 (S8-S5, S9-S3)	2 (S5-S3, S9-S8)

Z = w1(D1 + D2) + w2(B1 + B2) = 1(94+86) + 1(2+2) = 184

D = total distance traveled by the AGV due to the movements occurring during the entire operation of processing all the P parts.

The final optimal schedule for both the parts is shown in Figure 1.10.

The AGV movement for processing these parts according to their sequence on the optimal layout shown in Fig 1.9 is:

Part 1: $S_2 \rightarrow S_7 \rightarrow S_1 \rightarrow S_4 \rightarrow S_8 \rightarrow S_6 \rightarrow S_5 \rightarrow S_3 \rightarrow S_2$.
Part 2: $S_5 \rightarrow S_3 \rightarrow S_7 \rightarrow S_1 \rightarrow S_4 \rightarrow S_6 \rightarrow S_9 \rightarrow S_8 \rightarrow S_2$

The total distance D, number of backtracking B, and the value of objective function are shown in Table 1.10.

The optimal route is obtained in terms of the best tour in reasonable computation time. The computation time associated with the codes of the ACO Algorithm are of the order 0.251–3.118 sec.

1.8 CONCLUSIONS AND FUTURE SCOPE

From the study and research involved in the present work, the following conclusions are obtained:

1. The scheduling of jobs and machine assignments in a FMS should take care of part flexibility as well as operations flexibility.
2. This makes it a flexible job shop scheduling problem, which is highly complex in nature in terms of the computational resources involved.
3. AGV routing has to be an integral component of FMS scheduling.
4. The AGV routing problem forms a kind of Traveling Salesman Problem (TSP), which belongs to the class of NP complete problems and can be optimally solved only with the help of metaheuristics.
5. Ant Colony Optimization (ACO) can be a good choice for a single AGV routing problem. It is easy to implement and also capable of finding the optimal solution in terms of best tour in less computational time.

1.9 SCOPE FOR FURTHER WORK

The present work discusses a model of integrated scheduling and AGV routing in a FMS environment. In the considered problem case, a single AGV is used. However, in actual manufacturing facilities with a higher level of automation, there may be as many as a hundred slots for machine stations, and the parts are produced in batches. In such cases, multiple AGVs are used for material transport. This poses various other concerns that are not taken into consideration in this work. A few of these are:

- A layout plan for the unequal distribution of area in the manufacturing facility.
- Multiple AGVs, multiple tracks, and hence multiple routes.
- The optimal number of AGVs required in such a system.
- A network design of the AGVs in such a way as to avoid congestion.

Artificial intelligence and machine learning are the two important pillars over which a so-called factory of the future will stand tall. AI is constantly evolving. More advanced heuristics that mimic the swarm or flocking behavior of organisms like Grey Wolf Optimization (GWO) can be attempted for multiple AGV routing and design problems.

REFERENCES

[1] Groover, M.P. (1992). *Automation, Production Systems and Computer-integrated Manufacturing*. USA: Prentice-Hall Inc.

[2] Nambiar, A.N. (2010). Modern Manufacturing Paradigms – A Comparison. *Hong Kong. Proceedings of International Multi-Conference of Engineers and Computer Scientists*, Vol. III, IMECS 2010.

[3] Sethi, S.P., Stecke, K.E., Dubois, D., & Browne, J. (1984). Classification of flexible manufacturing systems. *The FMS Magazine*.

[4] Fazlollahtabar, H., & Saidi, M. *Autonomous Guided Vehicles, Studies in Systems, Decision and Control*. (Vol. 20). New York, Dordrecht, London: Springer Cham Heidelberg.

[5] Bedworth, D.D., & Bailey, J.E. (1987) *Integrated Production Control Systems*, 2nd edn., Singapore: Wiley.

[6] Stecke, K.E. (1983). Formulation and Solution of Nonlinear Integer Production Planning Problems for Flexible Manufacturing Systems. *Management Science*, v. 29, n. 3. (pp. 273–288).

[7] He, W., & Kusiak, A. (1992). Scheduling of manufacturing systems. *Computers in Industry* Vol. 20, Issue 2, (pp. 163–175).

[8] Hall, N.G., & Magazine, M. (2013). Scheduling and Sequencing, *Encyclopedia of Operations Research and Management Science*, (pp. 1356–136).

[9] Ioannou, G., & Kritikos, M.N. (2004). Optimization of material handling in production and warehousing facilities. *Oper Research, An International Journal.* Vol. 4, (pp. 317–331).

[10] Vosniako, G.C., & Mamalis, A.G. (1990). Automated guided vehicle system design for FMS applications. *Int. Journal of Machine Tools and Manufacture*, Vol, 30(1), (pp. 85–97).

[11] Fazlollahtabar, H., & Saidi-Mehrabad, M. (2013). Methodologies to Optimize Automated Guided Vehicle Scheduling and Routing Problems: A Review Study. *Journal of Intelligent & Robotic Systems*, v. 77(3–4), (pp. 525–545).

[12] Tetzlaff, A.W. (1990). *Optimal Design of Flexible Manufacturing System.*, Berlin Heidelberg: Springer-Verlag.

[13] Baker, K., & Trietsch, D. (2009). *Principles of Sequencing and Scheduling.* Hoboken, NJ: Wiley.

[14] Adams, J., Balas, E., & Zawack, D. (1988). The shifting bottleneck procedure for job shop scheduling. *Management Science*, v. 34, (pp. 391–401).

[15] Hansmann, K.W., & Michael-Hoeck. (1997). Production control of a flexible manufacturing system in a job shop environment. *International Transactions in Operational Research*, Vol. 4, Issues 5–6, (pp. 341–351).

[16] Ficko, M., Brezocnik, M., & Balic, J. (2004). Designing the layout of single- and multiple-rows flexible manufacturing system by genetic algorithms. *Journal of Materials Processing Technology*, Volumes 157–158, (pp. 150–158).

[17] Geiger, C.D., Uzsoy, R., & Aytuğ, H. (2006). Rapid Modeling and Discovery of Priority Dispatching Rules: An Autonomous Learning Approach. *Journal of Scheduling 9*, (pp. 7–34).

[18] Scheduling in Flexible Manufacturing Systems. *In: Handbook on Scheduling. International Handbook on Information Systems.* (2007). Berlin, Heidelberg: Springer.

[19] Kusiak, A., & Cyrus, J.P. (1985). Routing and Scheduling of Automated Guided Vehicles. In: Bullinger, H.J., & Warnecke H.J. (Ed.) *Toward the Factory of the Future.* Berlin, Heidelberg: Springer.

[20] Mahadevan, B., & Narendran, T.T. (1990). Design of an automated guided vehicle-based material handling system for a flexible manufacturing system. *International Journal of Production Research*, 28:9, (pp. 1611–1622).

[21] Krishnamurthy, N.N., Batta, R., and Karwan, M.H. (1991). Developing Conflict-Free Routes for Automated Guided Vehicle. *Working Paper, Department of Industrial Engineering, State University of New York at Buffalo.* Buffalo, NY.

[22] Li, Q., Adriaansen, A.C., Udding, J.T., & Pogromsky, A.Y. (2011). Design and Control of Automated Guided Vehicle Systems: A Case Study. *IFAC*, Vol 44, Issue 1, (pp. 13852–13857)

[23] Adulyasak Y, et al. (2014). The production routing problem: A review of formulations and solution algorithms. *Computers and Operations Research*.

[24] Mirjalili, S., Dong, J. S., & Lewis, A. (2020). Nature Inspired Optimizers. *Studies in Computational Intelligence*, Vol. 811, (pp. 7–20). Switzerland AG: Springer Nature.

[25] Chopard, B., & Tomassini, M. (2018). An introduction to metaheuristics for optimization. *Nature Computing Series*. Switzerland AG: Springer Nature.

[26] Kirkpatrick, S., Gelatt, C.D., & Vecchi. M.P. (1983). Optimization by simulated annealing. *Science*, 220(4598), (pp. 671–680).

[27] Kennedy, J., & Eberhart, J. (1995). Particle Swarm Optimization. *Proceedings of IEEE International Conference on Neural Networks. IV* (pp. 1942–1948).

[28] Mitchell, & Melanie. (1996). *An Introduction to Genetic Algorithms*. Cambridge, MA: MIT Press. ISBN 9780585030944.

[29] Dorigo, M., & Gamberdella, L.M. (1997). Ant colonies for the travelling salesman problem. *Bio Systems* (43) (pp. 73–81).

[30] Shi, Y., & Eberhart, R.C. (1998). A modified particle swarm optimizer. *Proceedings of IEEE International Conference on Evolutionary Computation*. (pp. 69–73).

[31] Kim C.W., & Tanchoco J.M.A. (1994). Bidirectional Automated Guided Vehicle Systems (AGVS). In: Tanchoco J.M.A. (Ed.) *Material Flow Systems in Manufacturing*. Boston, MA: Springer.

[32] Dorigo, M., & Di Caro, G. (1999). Ant colony optimization: a new metaheuristic. *Proceedings of the 1999 Congress on Evolutionary Computation-CEC99 (Cat. No. 99TH84060)*. (Vol. 2, pp. 1470–1477). Washington, DC, USA.

[33] Dorigo, M., Bonabeau, E., & Theraulaz, G. (2000). Ant algorithms and stigmergy. *Future Generation Computer Systems*. (16(8), pp. 851–871).

[34] Kumar, R., Tiwari, M. K., & Shankar, R. (2003). Scheduling of flexible manufacturing systems: An ant colony optimization approach. *Proc. Institution Mechanical Engineers, Vol. 217, Part B: J. Engineering Manufacture*.

[35] Sankar, S., Ponnanbalam, S.G., & Rajendran, C. (2003). A multiobjective genetic algorithm for scheduling a flexible manufacturing system. *Int J Adv Manuf Technol*, 22. (pp. 229–236).

[36] Jerald, J., Asokan, P., & Prabaharan, G. (2005). Scheduling optimisation of flexible manufacturing systems using particle swarm optimisation algorithm. *Int J Adv Manuf Technol* 25, (pp. 964–971).

[37] Gnanavel Babu, A., Jerald, J., Noorul Haq, A., Muthu Luxmi, V., & Vigneswaralu, T.P. (2010). Scheduling of machines and automated guided vehicles in FMS using differential evolution. *International Journal of Production Research*, 48:16, (pp. 4683–4699).

[38] Burnwal, S., & Deb, S. (2013). Scheduling optimization of flexible manufacturing system using cuckoo search-based approach. *Int J Adv Manuf Technol* 64, (pp. 951–959).

[39] Haq, A.N., Karthikeyan, T., & Dinesh, M. (2003). Scheduling decisions in FMS using a heuristic approach. *Int J Adv Manuf Technol*. 22. (pp. 374–379).

[40] Pratihar, D.K. Traditional vs. non-traditional optimization tools. Basu, K.. & Kar, S. (Ed.), *In book: Computational Optimization and Applications*, (pp. 25–33). New Delhi, IN: Narosa Publishing House Pvt. Ltd.

[41] MHI. (n.d.). Definition of Material Handling & Logistics. The Material Handling Institute of America. https://www.mhi.org/about.

Chapter 2

Effective energy management in smart buildings using VRV/VRF systems

Pardeep Kumar and Shalom Akhai

2.1 INTRODUCTION

We all know that in today's world, humanity is attempting to develop the most advanced technologies in order to improve personal comfort while also addressing new and upcoming survival challenges [1]. To achieve this, continuous developments and improvements are currently being made in a variety of processes and industries [2]. We have entered an era in which optimization and automation are critical to improving the efficiency of systems that are currently being developed or used in order to upgrade them [3].

Yet every process requires energy, and the vast use of the energy available on earth is resulting in global warming, which will be extremely harmful to humankind in the future [4]. The concept of sustainability, which is a major consideration in the modern state of industry, is a major challenge in Industry 4.0 [5]. Devices such as air conditioners that are used for human comfort [6] and are used in today's large buildings to create a comfortable environment for those who live and work inside them [7]. Several studies have shown that air conditioners consume a significant amount of electricity in any building, particularly during the summer months, accounting for nearly 60% of all electricity consumed by air conditioning systems [8]. Air conditioning systems, such as those found in hotels, hospitals, shopping malls, and commercial buildings, rely on a vapor compression refrigeration system [9–10]. Vapor absorption systems are commonly used in industrial settings and buildings [11]. In today's world, the vapor compression refrigeration system uses a technology that is geared toward sustainability in terms of the amount of energy consumed by the air conditioner while in use [12]. Variable refrigerant volume (VRV) or variable refrigerant flow (VRF) is a technology developed by large air conditioning companies like

DOI: 10.1201/9781003360001-2

Daikin [13]. However, many important and historic buildings have not been retrofitted with this latest technology, which, even though new, has proven to be effective in achieving energy efficiency and green certifications for the buildings in which it has been installed [14]. Many technical societies, such as ASHRAE (American Society of Heating, Refrigeration, and Air Conditioning) and ISHRAE (Indian Society of Heating, Refrigeration, and Air Conditioning), have been actively promoting this technology in recent years, with which, as a result, a building can easily obtain certification for green infrastructure and sustainability [15]. In the case of residential air conditioning, the conventional window and split air conditioners are widely used for human comfort [16]. However, central systems are preferred in the context of the potential for virus spread over conventional window and split air conditioners [17]. In central systems the VRV/VRF technology has been developed after much of research and a number of improvements since the advent of the VRV system some decades ago. Due to the rapid growth of the Indian economy, a developing country like India still has a poor supply of energy compared to its growing demands, and this supply–demand gap is widening all the time, as evidenced by the fact that energy demand has increased by more than 3.6% per year over the last 30 years [18].

Why, then, is the world turning to VRV? There are four main reasons [19]:

1. *Energy saving:* Higher COP and VRT (variable refrigerant temperature)
2. *Ease of installation:* Compact and lightweight design
3. *Comfort:* Lower operating sound
4. *Enhanced lineup:* Up to 60 HP

VRT (variable refrigerant temperature) is the latest in VRV/VRFF technology. VRT automatically adjusts refrigerant temperature to individual building and climate requirements, thus further improving annual energy efficiency and maintaining comfort. With this technology, running costs are reduced [19]. See Figure 2.1.

When cooling, the refrigerant evaporating temperature is raised to minimize the difference with the condensing temperature. When heating, the condensing temperature is lowered to minimize the difference from the evaporating temperature. So the compressors work less, and this reduces power consumption [20]. With VRV technology, there are indoor units and outdoor units, as always in every other type of air conditioner. The indoor units are well equipped with sensors, which detect the change in temperature and the load on the system. The occupancy in a particular room is detected by the sensors on the indoor unit, and the airflow direction is altered as per the sensor results [21]. The temperature near each person is calculated from the floor temperature, and the air flow direction and the load are changed according to requirements. As the occupancy for a particular indoor unit

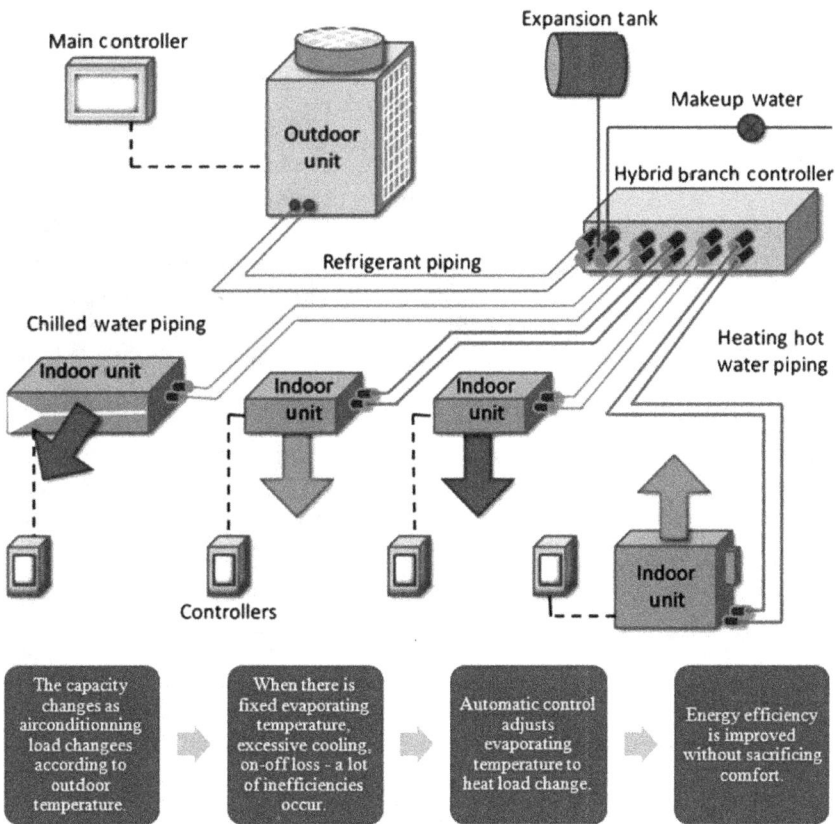

Figure 2.1 Schematic sequencing operation in VCRS system.

increases, the sensor detects the increase of load. This sends a message to the control data center, and this increases the refrigerant flowing to the indoor unit. The cooling in that particular room increases, thereby providing comfort conditions [22]. As occupancy decreases, the sensor detects the fall in load, and this leads to a decrease in the quantity of refrigerant flowing to the indoor unit [23]. The outdoor unit has a great lineup option. The outdoor unit of the VRV has the ability to check the wiring and the pipe connections all by itself. Any breaks and leakages can thereby be analyzed by the system itself. This helps in reducing the maintenance time and the service cost. The outdoor sequencing is also altered so that the starting load does not always fall on the same compressor over and over again. This improves the lifetime of the compressor [22].

2.2 LITERATURE REVIEW

The primary purpose of an HVAC system is cooling and providing thermal comfort, and for the central systems that serve these purposes, VRV/VRF technology was researched and developed a few decades ago [23]. According to the literature, many studies have been conducted showing that these systems are cost-effective and reduce electricity bills by increasing energy efficiency [24]. A lot of work is also being done to improve these systems by modifying them to accommodate buildings that are located at different latitudes and longitudes, as well as at geographic locations that differ from one another [25]. Because each building's heat load is calculated separately, the size and design of an air conditioning system that is appropriate for that particular building can be determined. As result, air conditioning design and location vary, so additional modifications may be made by integrating minor modifications to improve efficiency even further [26]. An analysis of the past studies would be helpful in understanding the gaps existing in such system and ways to overcome them, thereby providing humankind with the means to meet requirements in a more economical and optimized manner. A few advances in this direction are discussed in Table 2.1.

2.3 CONCLUSION

In conclusion, we can say that many explorations are going on in the field of air conditioning technology, particularly in the latest technology of variable refrigerant volume and variable refrigerant flow, because air conditioners are used all over the world, and different buildings have varying orientations and climatic conditions. So each system must be optimized based on the heat load calculation and building requirements. After an overall study of the technology, it can be said with certainty that this technology is the need of the hour. The system has the ability to save a great deal of energy when compared with the constantly running air conditioners. Such system have the proven reputation of being an energy saving option and deployable in almost all applications. The VRV/F system can be optimized as per the need and therefore has a lot of flexibility in terms of output characteristics and performance. In future studies, research can be conducted in the field of refrigerant usage, so as to have environmentally friendly refrigerants. The system shows excellent results when used on part loads therefore more studies need to be done on full loads. The full load of the VRV systems needs to be developed so as to get an improved energy performance characteristic. As for the current scenario, the only area where the VRV system is lacking is the full load performance. When it comes to full load performance, the behavior of the system is equivalent to the constant volume system; therefore it is necessary to work on the optimization of energy savings on full load so as to get a completely energy efficient system.

Table 2.1 Advances in VRV/VRF

Source	Research field	Component	Results	Gaps
Zhu et al. 2015 [27]	Tried to introduce the concept of combining both variable refrigerant flow and the variable air volume, so as to achieve higher efficiency.	The research based on improving the efficiency of the system full load by the joint application of VRF and VAV.	The combined system showed effective reduction in energy consumption, thereby accomplishing the objective of the study.	The system lowered the energy consumption of the building space, but there was not a full -scale thermal comfort as experienced in the individual system having VRF systems standalone.
D. Zhao, Zhang, and Zhong 2015 [28]	The study was based on improving ways of refrigerant flow so as to obtain the highest efficiency possible using the available resources.	The study was conducted to reveal the basic differences between the part- load applications and the full - load applications	The results show that the evaporating temperature rises and the energy consumption decreases at a gradual pace.	The difference between the evaporating temperature in the heat exchanger and the saturated temperature needs to be improved so as to get higher successful rates of energy efficiency.
Yu et al. 2016 [29]	The study was based on finding the comparison between the variable air volume and the variable refrigerant flow systems.	The study aimed at analyzing the energy consumption at an office building in China.	The data showed that the VRF system used up to 70% lower energy as compared to the VAV system.	The cooling load of the system was lowered by 42%, which showed the potential of going even lower with proper research. The ability to design net-zero buildings using VRF systems needed to be developed.
D. Zhao et al. 2016 [30]	The study targets at studying the energy performance characteristics of the VRV system based on the energy consumption inter sity (ECI).	The ECI is applied to various office buildings in Shanghai and other cities in mainland china so as to get an idea of the energy consumption of the existing VRV system in the offices.	It is found that the practical energy saving limits of the VRV systems are at par with the theoretical calculations of the ECI.	The ANN and SVM models may have to do energy consumption calculations of the buildings so as to get better results.

(continued)

Table 2.1 Cont.

Source	Research field	Component	Results	Gaps
Al-Aifan et al. 2017 [31]	The study was targeted at combining the variable refrigerant volume (VRV) and cool thermal energy storage (CTES) so as to get proper summer and winter load conditions.	This system was primarily used in the storage of food items and other cold storage applications. The VRV CTES can maintain the indoor temperature at an almost constant rate.	The VRV CTES system precisely maintained the indoor temperature at 24°C throughout the entire year.	It would be beneficial for this system to provide comfort conditions by meeting the validated maximum energy consumptions so as to be a more environmentally friendly system.
Khatri and Joshi 2017 [32]	The author aimed at studying the relation between the variable refrigerant system and the constant volume system to find the more efficient system.	The research was based on the energy consumption by the air-conditioning system.	It was seen that the VRF system was more efficient as compared to the constant volume system.	The VRF was efficient only in the case of part load, and full load on VRF system was using energy equivalent to the constant system. More developments were needed to improve the performance of the maximum load on VRV system.
L Zhao et al. 2017 [33]	Using the variable refrigerant flow system for small to medium-sized office spaces so as to replace the present temperature humidity independent control (THIC) system.	The VRF system was used to dehumidify the outdoor air and also for refrigerant water plate heat exchanger system.	The new novel system developed in this system showed appreciating performance in regard to the temperature and humidity control.	The ratio of energy performance improvement and the effect on the indoor environment required some more detailed study, which was not part of the present study.
Kim et al. 2017 [34]	The software EnergyPlus has been used to study the performance of the VRF and RTU VAV systems in a small office space.	The study was systematically carried out in 16 different places so as to get a characteristic idea of all the different places with varying climatic conditions.	The result shows that the hot and mild places showed more energy saving as compared to the hot places.	There needs to be an improvement in the system designed for cold climatic places, because the system at such places showed really low energy savings mainly because of the additional use of gas for heating purposes.
Wei et al. 2020 [35]	Tried to apply the use of VRF system in heat pumps currently operating in regions with extremely low temperature conditions.	A test rig was set up in Harbin, and the VRF system of heat pump was used to heat up the laboratory.	The unit was able to provide continuous heating even when the outdoor temperature was −22°C.	The heating capacity and the discharge temperature were all positives for the system, while mal defrost was affecting the heat storage capacity of the system.

REFERENCES

[1] Diamandis, P. H., & Kotler, S. (2020). *The future is faster than you think: How converging technologies are transforming business, industries, and our lives.* Simon & Schuster.

[2] Muñoz-La Rivera, F., Mora-Serrano, J., Valero, I., & Oñate, E. (2021). Methodological-technological framework for Construction 4.0. *Archives of Computational Methods in Engineering, 28*(2), 689–711.

[3] Javaid, M., Haleem, A., Singh, R. P., & Suman, R. (2021). Significance of Quality 4.0 towards comprehensive enhancement in manufacturing sector. *Sensors International, 2*, 100109.

[4] Malhi, Y., Franklin, J., Seddon, N., Solan, M., Turner, M. G., Field, C. B., & Knowlton, N. (2020). Climate change and ecosystems: Threats, opportunities and solutions. *Philosophical Transactions of the Royal Society B, 375*(1794), 20190104.

[5] Beier, G., Ullrich, A., Niehoff, S., Reißig, M., & Habich, M. (2020). Industry 4.0: How it is defined from a sociotechnical perspective and how much sustainability it includes–A literature review. *Journal of Cleaner Production, 259*, 120856.

[6] Akhai, S., Singh, V. P., & John, S. (2016). Investigating Indoor Air Quality for the Split-Type Air Conditioners in an Office Environment and Its Effect on Human Performance. *Journal of Mechanical Civil Engineering, 13*(6), 113–118.

[7] Tanwar, N., & Akhai, S. (2017). Survey Analysis for Quality Control Comfort Management in Air Conditioned Classroom. *Journal of Advanced Research in Civil and Environmental Engineering, 4*(1&2), 20–23.

[8] Akhai, S., Bansal, S. A., & Singh, S. (2020). A critical review of thermal insulators from natural materials for energy saving in buildings. *Journal of Critical Reviews, 7*(19), 278–283.

[9] Elsaid, A. M., & Ahmed, M. S. (2021). Indoor Air Quality Strategies for Air-Conditioning and Ventilation Systems with the Spread of the Global Coronavirus (COVID-19) Epidemic: Improvements and Recommendations. *Environmental Research*, 111314.

[10] Akhai, S., Mala, S., & Jerin, A. A. (2021). Understanding whether Air Filtration from Air Conditioners Reduces the Probability of Virus Transmission in the Environment. *Journal of Advanced Research in Medical Science & Technology (ISSN: 2394-6539), 8*(1), 36–41.

[11] Choudhury, B., Saha, B. B., Chatterjee, P. K., & Sarkar, J. P. (2013). An overview of developments in adsorption refrigeration systems towards a sustainable way of cooling. *Applied Energy, 104*, 554–567.

[12] Papadopoulos, A. M., Oxizidis, S., & Kyriakis, N. (2003). Perspectives of solar cooling in view of the developments in the air-conditioning sector. *Renewable and Sustainable Energy Reviews, 7*(5), 419–438.

[13] Wang, X., Xia, J., Zhang, X., Shiochi, S., Peng, C., & Jiang, Y. (2009, July). Modelling and experiment analysis of variable refrigerant flow air-conditioning systems. In *Proceedings of the IBPSA conference on building simulation* (pp. 361–368).

[14] Duggin, C. (2018). VRF: Overcoming challenges to achieve high efficiency: Variable refrigerant flow systems can result in effective, high-efficiency HVAC designs, but care must be taken to achieve proper ventilation, humidity control, and compliance with ASHRAE Standard 15. *Consulting Specifying Engineer, 55*(3), 12–16.

[15] Sharma, R., Raman, R., Yadav, O., Khan, M. S., & Yadav, V. K. (2022). Thermal Comfort and Energy Efficient Design of a Central Air Conditioning System of an Educational Institute. In *Recent Trends in Thermal Engineering* (pp. 143–159). Springer, Singapore.

[16] Arsana, M. E., Kusuma, I. G. B. W., Sucipta, M., & Suamir, I. N. (2021). Exergy and Energy Analyses of Dual-Temperature Evaporator Split AC System Incorporated A Capillary Tube and A Two-Phase Ejector. *Journal of Advanced Research in Fluid Mechanics and Thermal Sciences, 77*(1), 88–99.

[17] Akhai, S., Mala, S., & Jerin, A. A. Apprehending Air Conditioning Systems in Context to COVID-19 and Human Health: A Brief Communication. *International Journal of Healthcare Education & Medical Informatics* (ISSN: 2455–9199). 2020;7(1&2).

[18] Kumar, P., Tewari, P., & Khanduja, D. (2017). Six Sigma application in a process industry for capacity waste reduction: a case study. *Management Science Letters, 7*(9), 423–430.

[19] Aynur, T. N. Variable refrigerant flow systems: A review. *Energy and Buildings.* 2010 Jul 1;42(7):1106–12.

[20] Ahamed, J. U., Saidur, R., Masjuki, H. H. A review on exergy analysis of vapor compression refrigeration system. *Renewable and Sustainable Energy Reviews.* 2011 Apr 1;15(3):1593–600.

[21] Chuang, H. C., Zeng, Y. X., & Lee, C. T. Study on a chiller of air conditioning system by sensing refrigerant pressure feedback control with stepless variable speed driving technology. *Building and Environment.* 2019 Feb 1;149:157–68.

[22] Ali, S. B. M., Hasanuzzaman, M., Rahim, N. A., Mamun, M. A. A., & Obaidellah, U. H. (2021). Analysis of energy consumption and potential energy savings of an institutional building in Malaysia. *Alexandria Engineering Journal, 60*(1), 805–820. Aynur, T. N., Hwang, Y., & Radermacher, R. (2008). Experimental evaluation of the ventilation effect on the performance of a VRV system in cooling mode – Part I: Experimental evaluation. *HVAC&R Research, 14*(4), 615–630.

[23] Ozbalta, T. G., Yildiz, Y., Bayram, I., & Yilmaz, O. C. (2021). Energy performance analysis of a historical building using cost-optimal assessment. *Energy and Buildings, 250*, 111301.

[24] Nayak, B. K., & Rajan, P. (2021). The Crossroads on the Path to Sustainability While Aspiring for a Better Quality of Life: A Case of Delhi. In *Handbook of Quality of Life and Sustainability* (pp. 521–531). Springer, Cham.

[25] Ananda, M. H., Raghavendra, K., Ballaji, A., Ankaiah, B., Sagar, B. S., Raghu, C. N., & Doddabasappa, N. (2021, March). Techno-economic assessment of air cooling/ventilating methods for the college convention center. In *IOP Conference Series: Materials Science and Engineering* (Vol. 1114, No. 1, p. 012032). IOP Publishing.

[26] Kanaan, A., Sevostianova, E., Donaldson, B., & Sevostianov, I. (2021). Effect of Different Landscapes on Heat Load to Buildings. *Land*, *10*(7), 733.

[27] Zhu, Y., Jin, X., Du, Z., & Fang, X. (2015). Online optimal control of variable refrigerant flow and variable air volume combined air conditioning system for energy saving. *Applied Thermal Engineering*, *80*, 87–96.

[28] Zhao, D., Zhang, X., & Zhong, M. (2015). Variable evaporating temperature control strategy for VRV system under part load conditions in cooling mode. *Energy and Buildings*, *91*, 180–186.

[29] Yu, X., Yan, D., Sun, K., Hong, T., & Zhu, D. (2016). Comparative study of the cooling energy performance of variable refrigerant flow systems and variable air volume systems in office buildings. *Applied Energy*, *183*, 725–736.

[30] Zhao, D., Zhong, M., Zhang, X., & Su, X. (2016). Energy consumption predicting model of VRV (Variable refrigerant volume) system in office buildings based on data mining. *Energy*, *102*, 660–668.

[31] Al-Aifan, B., Parameshwaran, R., Mehta, K., & Karunakaran, R. (2017). Performance evaluation of a combined variable refrigerant volume and cool thermal energy storage system for air conditioning applications. *International Journal of Refrigeration*, *76*, 271–295.

[32] Khatri, R., & Joshi, A. (2017). Energy performance comparison of inverter based variable refrigerant flow unitary AC with constant volume unitary AC. *Energy Procedia*, *109*, 18–26.

[33] Zhao, L., Jianbo, C., Haizhao, Y., & Lingchuang, C. (2017). The development and experimental performance evaluation on a novel household variable refrigerant flow based temperature humidity independently controlled radiant air conditioning system. *Applied Thermal Engineering*, *122*, 245–252.

[34] Kim, D., Cox, S. J., Cho, H., & Im, P. (2017). Evaluation of energy savings potential of variable refrigerant flow (VRF) from variable air volume (VAV) in the US climate locations. *Energy Reports*, *3*, 85–93.

[35] Wei, W., Ni, L., Xu, L., Yang, Y., & Yao, Y. (2020). Application characteristics of variable refrigerant flow heat pump system with vapor injection in severe cold region. *Energy and Buildings*, *211*, 109798.

Chapter 3

Kinematic analysis, modeling, and simulation of a 7-link biped robot

Arun Kumar Kushwaha, Ram Bilas Prasad, and Darshan Srivastav

DOI: 10.1201/9781003360001-3

3.1 INTRODUCTION

Humans have always been interested in learning from nature and try to mimic the activities happening all about their surroundings. A humanoid robot is the result of this human curiosity. Since the humanoid robot is a complex thing, a biped (two-legged) robot is needed to simplify it. Humans and some other animals like birds and kangaroos have bipedal locomotion.

Since the study of whole-body motion is a complicated task, to easily analyze and describe bipedal locomotion, a biped robot is used. The biped robot is a simple version of a humanoid robot in which the upper body is neglected. The biped robot has many advantages compared to wheeled robots. It provides greater mobility. Biped robots find significant application in the development of prosthetics for physically challenged persons who are unable to walk. Biped locomotion also plays an important role in making an exoskeleton to support elderly people and assist military personnel or others who are supposed to stand or walk with heavy load for a prolonged duration.

A biped robot is a simplified version of humanoid robot that mimics human motion. A biped robot generally consists of legs, a foot, and a torso, which provide simplicity in bipedal locomotion. S biped robot gets much attention for its ability to walk over flat surfaces and uneven surfaces like stairs, rough crossings, ditches, etc., which wheeled robots are not capable of doing. .The biped robot replaces the human when the work poses health risks, like mining, military, space, etc. It may also be used in helping society in pandemic situation where work may be done without direct human involvement.

3.1.1 Related works

Gupta and Dutta [1] presented a kinematic and dynamic model of a biped that implements a joint trajectory to avoid obstacles during walking. A neural network approach for optimization of the energy during walking was also attempted. A camera capture experiment was performed by Ankarali [2] in order to evaluate the angular velocities of various joints, which worked as input to the fuzzy logic controller. The output of the fuzzy logic controller was used to control the servomotor, and the response to the servomotor worked as input to the SimMechanics model of the biped to generate gait motion.

Sarkar and Dutta [3] presented a finite-element-based approach for modeling four different shapes of compliant shank in an 8-degrees-of-freedom biped robot. Genetic Algorithm was used for optimizing the trajectory for minimum energy consumption during walking. A model of a shape memory alloy-based actuator in the biped was presented by Moghadaam et al. [4], with the advantages such as noiseless motion and low weight-to-flexibility ratio. Vundavilli and Pratihar [5] proposed the genetic-neural and genetic-fuzzy logic control method to optimize the gait of the biped during ascending and descending on a slope. They analyzed that power consumption during ascending was greater than on descending. Lu et al. [6] proposed the turning base behavior of biped. Three different approaches were used for comparison: body *first* method for instant turning, *leg fist* method when space behind the leg is less, and simultaneous body leg movement when a large turn angle is required. Sun et al. [7] proposed the neural network for control wherein a radially based function was used for approximating some unknown parameters. Various conditions of the biped, such as ascending and descending, were simulated and controlled by Mousavi and Bagheri [8]. Two types of ZMP (zero moment point) – fixed and moving – were used. Also, computation of various kinematic and dynamic parameters like inertia force and joint angle was presented.

Panwar and Sukavanam [9] proposed a neural network method for solving the inverse kinematics problem of a biped robot that was able to walk. They also studied the effect of upper body movement instability. They assumed that in the case of the swing leg, the hip works as base and the foot as an end effector and that, in the case of the stable leg, the foot works as a base and the hip as an end effector. Mandava and Vundavilli [10] proposed the MCIW Algorithm for optimizing the gain in the PID (proportional–integral–derivative) controller of the motor. They tuned the controller manually and compared the MCIW optimization with differential evolution optimization for comparing various biped robot properties like error, ZMP, need of torque, DBM (dynamic balance margin). They find the MCIW gives a better result. Sugahara et al. [11] used Stewart Platforms as leg mechanism of the biped robot, which was experimented with in ascending and descending stairs. Lin et al. [12] developed a flexible ankle joint with a feedback linearization method and a virtual constrained method controlling joint angles. Prasad and Arif [14] proposed inverse kinematics and a Jacobian approach for finding workspace and singularity of a five-link parallel mechanism.

3.2 METHODS

3.2.1 Kinematic analysis

Kinematics is the study of motion only, not the cause behind the motion. In kinematics, the motion of different links, relative motion between links, acceleration, velocity, etc. are analyzed without studying the torques and forces that cause the motion of the links.

Robots or manipulators have many links connected to one another and having a different frame. It is important to find the position and orientation of links concerning the base frame. So the study of kinematics is essential for analyzing the motion.

There are two modes of kinematics used in robotics: forward kinematics and inverse kinematics. Before going on to the kinematic analysis of a robot, it is very important to know about the plane of analysis.

3.2.2 Planes of analysis

There are three types of plane for studying the humanoid robot motion, as shown in Figure 3.1.

1. *Sagittal plane:* The longitudinal plane that divides the body into the left and right parts

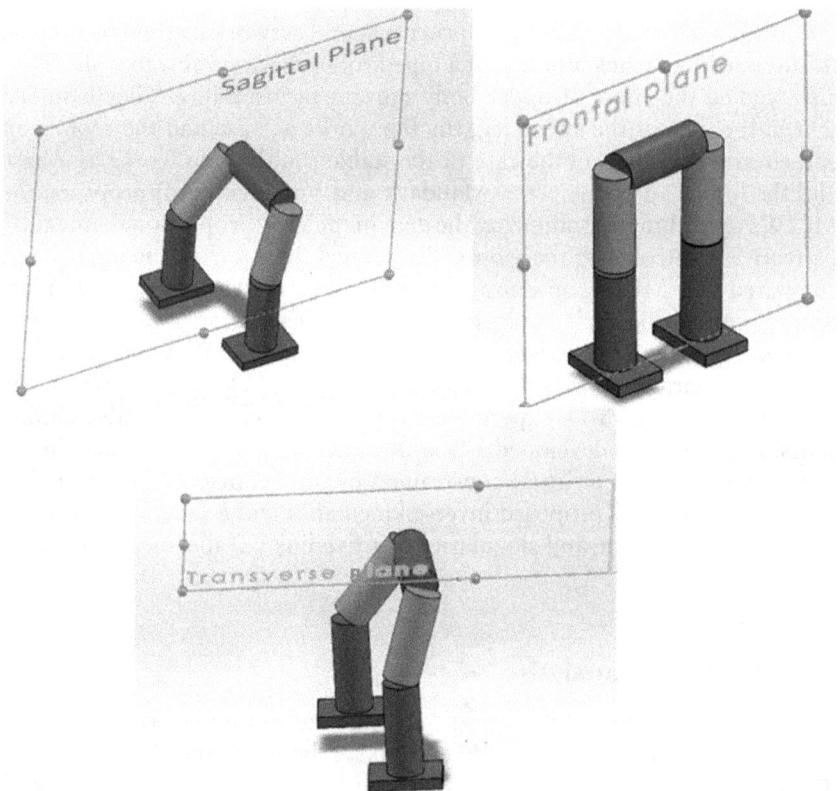

Figure 3.1 Plane of Analysis.

2. *Frontal plane:* The plane that is perpendicular to the longitudinal plane dividing the body into the front and back portions

3. *Transverse plane:* The plane that is perpendicular to both planes and that divides the body into the upper and lower portions

3.2.3 Walking cycle

There are two types of modes in bipedal motion, as shown in Figure 3.2. The first is the *single support phase* (SSP), and another is the *double support phase* (DSP). The single support phase is the mode of walking in which one leg is on the ground, and the other one is in the air. In DSP, both legs are on the ground. A walking cycle consists of two single support phases and two double support phases. In normal walking conditions, approximately 20% of the walking cycle time is consumed during DSP and approximately 80% during SSP. During running, it is observed that only one leg is on the ground at any time in bipedal locomotion. So, during running, only SSPs exist.

In the analysis of a biped robot, the gait period is defined as time taken by one leg to complete one stance phase and one swing phase.

3.2.4 Forward kinematics

Forward kinematics is the method of finding the position of the robot with the given joint parameters. The joint angles are the known parameters in this method, and the position of the robot is computed.

DH parameter of biped robot

The DH parameter concept was proposed by Denavit Hartenberg in 1955. In this concept, the coordinate system can be assigned easily for different joints, so that one can find the position and orientation of different joints concerning the base coordinate system. Figure 3.3 shows a 7-link and 6-joint biped robot in which the foot is connected to lower leg with the ankle

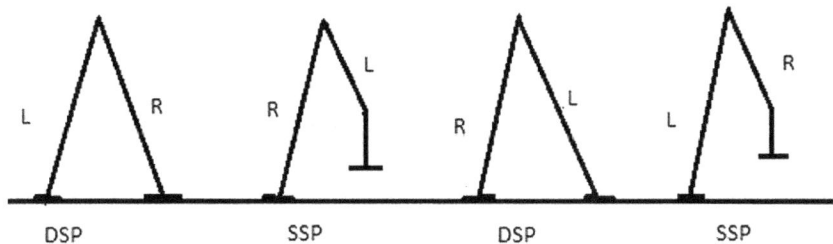

Figure 3.2 Walking cycle.

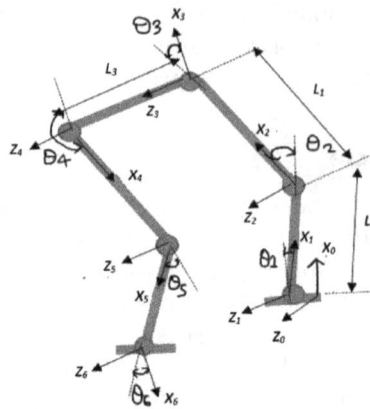

Figure 3.3 Biped robot.

Table 3.1 DH parameters of biped robot

θ_i	d_i	α_i	a_i
θ_1	0	0	0
θ_2	0	0	L_2
θ_3	0	0	L_1
θ_4	L_3	0	0
θ_5	0	0	L_1
θ_6	0	0	L_2

joint. The lower leg of the biped is connected to the upper leg with the knee joint. The upper leg is connected to the torso with the hip joint. The present work focuses on the study of the motion of the biped robot that involves only the pitching action of ankle, knee, and hip joint as all the joints are considered as revolute joints. To determine the position and orientation of the link, it is necessary to assign the coordinates to the joints according to the Denavit–Hartenberg (D-H) convention.

The notation of the DH parameter is as follows:

Link length (a$_i$): The distance between Z_{i-1} and Z_i in the direction of X_i
Angle of twist (α$_i$): The angle between Z_{i-1} and Z_i in the direction of X_i
Link offset (d$_i$): The distance between X_{i-1} and X_i in the direction of Z_{i-1}
Joint angle (θ$_i$): The angle between X_{i-1} and X_i in the direction of Z_{i-1}

Homogeneous transformation matrices of biped robot

Homogeneous transformation is needed when the position and orientation from one coordinate frame are transformed into another coordinate frame. In the transformation matrix $^{i-1}T_i$ denotes that the transformation has been taken from the i coordinate frame to the $i-1$ coordinate frame. The transformation matrix $^{i-1}T_i$ is determined using D-H parameter table.

$$^{i-1}T_i = \text{Rotation}(Z, \theta_i) * \text{Translation}(Z, d_i) * \text{Rotation}(X, \alpha_i) *$$
$$\text{Translation}(X, a_i) \tag{3.1}$$

$$^{i-1}T_i = (\text{screw } Z) * (\text{screw } X) \tag{3.2}$$

$$^{i-1}T_i = \begin{bmatrix} \cos\theta i & -\sin\theta_i\cos\alpha_i & \sin\theta_i & a_i \\ \sin\theta_i & \cos\theta_i\cos\alpha_i & \cos\theta_i\sin\alpha_i & a_i\sin\theta_i \\ 0 & \sin\alpha_i & \cos\alpha_i & d_i \\ 0 & 0 & 0 & 1 \end{bmatrix}$$

$$^0T_1 = \begin{bmatrix} 1 & 0 & 0 & 0 \\ 0 & 1 & 1 & 0 \\ 0 & 0 & 1 & d_1 \\ 0 & 0 & 0 & 1 \end{bmatrix}$$

$$^1T_2 = \begin{bmatrix} \cos\theta_2 & -\sin\theta_2 & \sin\theta_2 & L_2 \\ \sin\theta_2 & \cos\theta_2 & 0 & L_2\sin\theta_2 \\ 0 & 0 & 1 & 0 \\ 0 & 0 & 0 & 1 \end{bmatrix}$$

$$^2T_3 = \begin{bmatrix} \cos\theta_3 & -\sin\theta_3 & \sin\theta_3 & L_1 \\ \sin\theta_3 & \cos\theta_3 & 0 & L_1\sin\theta_3 \\ 0 & 0 & 1 & 0 \\ 0 & 0 & 0 & 1 \end{bmatrix}$$

$$^3T_4 = \begin{bmatrix} \cos\theta_4 & -\sin\theta_4 & \sin\theta_4 & 0 \\ \sin\theta_4 & \cos\theta_4 & 0 & 0 \\ 0 & 0 & 1 & L_3 \\ 0 & 0 & 0 & 1 \end{bmatrix}$$

$$
{}^4T_5 = \begin{bmatrix} \cos\theta_5 & -\sin\theta_5 & \sin\theta_5 & 0 \\ \sin\theta_5 & \cos\theta_5 & 0 & L_1\sin\theta_5 \\ 0 & 0 & 1 & 0 \\ 0 & 0 & 0 & 1 \end{bmatrix}
$$

$$
{}^5T_6 = \begin{bmatrix} \cos\theta_6 & -\sin\theta_6 & \sin\theta_6 & L_2 \\ \sin\theta_6 & \cos\theta_6 & 0 & L_2\sin\theta_5 \\ 0 & 0 & 1 & 0 \\ 0 & 0 & 0 & 1 \end{bmatrix}
$$

The homogeneous transformation of position and orientation of link in coordinate frame 6 to coordinate frame 0 is shown as follows:

$$
{}^0T_6 = {}^0T_1\,{}^1T_2\,{}^2T_3\,T_4\,{}^4T_5\,{}^5T_6 \tag{3.3}
$$

3.2.5 Inverse kinematics of the biped robot

The biped robot has seven links and six joints, and each leg has three revolute joints. The inverse kinematics is needed for finding the joint variables of a biped robot. Before finding the joint variables of the biped, it is important to know the position of the leg to apply inverse kinematics.

Finding positions X and Z

There are two positions of legs at the time of bipedal walking. One leg is stable, and the other leg swings while walking.

For the swing leg

It is assumed that the Z-coordinate of the biped swing leg makes a cubic polynomial with the X- coordinate and that the X-coordinate of the biped has a linear relation with time.

$$
x = a_0 + a_1 t \tag{3.4}
$$

where a_0, a_1 are constant, t is time, t_f is half the gait period, and x_f is the step length.

Boundary conditions: At $t = 0$, $x = 0$
At $t = t_f$, $x = x_f$

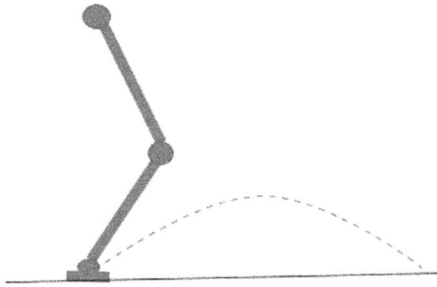

Figure 3.4 Swing leg trajectory.

Figure 3.4 shows that the walking direction of swing leg makes a cubic polynomial with respect to height during walk.

$$z(x) = b_0 + b_1 x + b_2 x^2 + b_3 x^3 \tag{3.5}$$

Boundary conditions: At $z = 0$, $x = 0$
At $z = h$, $x = (x_f)/2$ and $z = 0$, $x = x_f$

For the stable leg
In the biped simulation, the torso is connected to 6-degrees-of-freedom joint so that it is able to move in all directions. This condition is useful when the swing leg moves the torso of the biped, adjusting itself automatically because it is connected to the 6-degrees-of-freedom joint.

Finding angles

In this work, one leg (upper and lower leg) of the biped is considered a 2-R serial manipulator, and other leg a 2-R serial manipulator. Θ_1, Θ_2, Θ_3 are considered the hip joint angle, knee joint angle, and ankle joint angle, respectively.

Approach for determining hip joint and knee joint angle

Here assume that two links of length L_1 and L_2 are connected to the revolute joint. Link L_1 is connected with the base with angle θ_1 with respect to the vertical axis, and L_2 is connected with link 1 with angle θ_2, as shown in Figure 3.5.

$$X = L_1 \sin \theta_1 + L_2 \sin(\theta_1 + \theta_2) \tag{3.6}$$

$$Z = -L_1 \cos \theta_1 - L_2 \cos(\theta_1 + \theta_2) \tag{3.7}$$

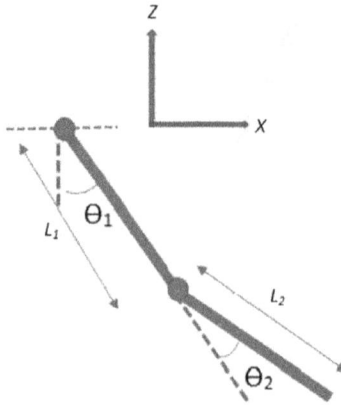

Figure 3.5 Upper leg and Lower leg as a links of manipulator.

By solving equations (3.6) and (3.7), the joint variable θ_1 and θ_2 can be determined as:

$$\theta_2 = \cos^{-1}\left(\frac{X^2 + X^2 - L_1^2 - L_2^2}{2L_1 L_2}\right) \tag{3.8}$$

$$\theta_1 = \tan^{-1}\left(\frac{X(L_1 + L_2 \cos\theta_2) + Z(L_2 + \sin\theta_2)}{X(L_2 \sin\theta_2) - Z(L_1 + L_2 \cos\theta_2)}\right) \tag{3.9}$$

Approach for determining ankle joint angle

In this analysis of biped robot, the foot is assumed to be always in a horizontal direction during the motion, and, during modeling, it is perpendicular to the lower leg, as shown in Figure 3.6. So for calculating the ankle joint based on this approach:

$$\theta_3 = 2\pi - (\theta_1 + \theta_2) \tag{3.10}$$

3.3 STABILITY

The concept of stability is very important for the analysis of walking of the biped robot. The following section discusses the concept of zero movement point (ZMP), which is an important parameter for finding the stability of the biped robot. The terminologies of the convex set, the convex and support polygon are defined to introduce the concept of ZMP.

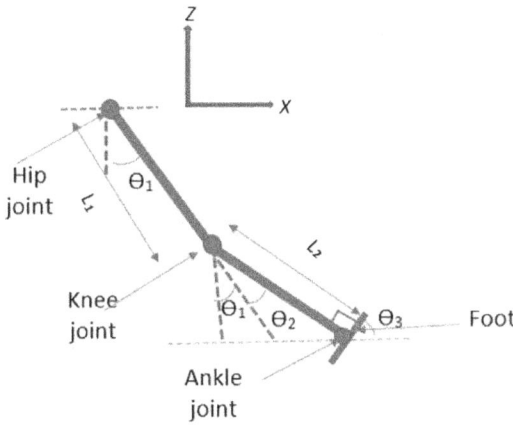

Figure 3.6 Ankle joint calculation.

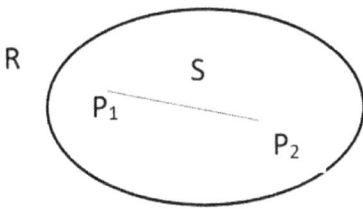

Figure 3.7 Convex set.

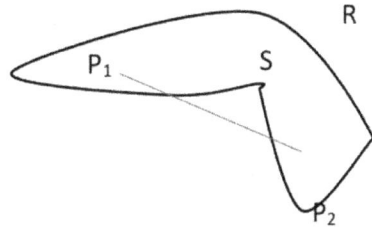

Figure 3.8 Non-convex set.

3.3.1 Convex set

If the line segment joining any two points of the set lies completely inside s, then s is called the *convex set*. Figures 3.7 and 3.8 represent a convex set and nonconvex set.

Let R be the region in which S is the subset. The subset S is convex set if:

$$\alpha P_1 + (1 - \alpha)P_2 \in S \tag{3.11}$$

where P_1 and P_2 are the points for any $P_1, P_2, \in S$, and the value of α is $0 \le \alpha \le 1$.

3.3.2 Convex hull

The convex hull is the polygon that contains the smallest convex set or polygon that encloses all points in the set, as shown in Figure 3.9. In the following sections, some algorithms for finding a convex hull are presented.

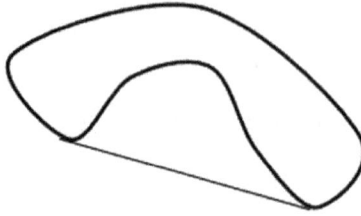

Figure 3.9 Convex hull.

Graham Scan Algorithm

The steps for the Graham Scan Algorithm are as follows:

1. Select the lowest *y*-coordinate.
2. Sort the points by the angles relative to the bottom mesh point and horizontal.
3. Iterate in sorted order, placing each point on a stack but only if it makes a counterclockwise turn relative to the previous two points on the stack.
4. Pop the previous point off the stack if making a clockwise turn.

In the Figure: 3.10, the first graphic (upper left) is of the points for finding the convex hull. In the second picture, we select the point that is the lowest-most of the vertical, i.e. point 0. The angle between point 0 and 1 is the minimum in the counterclockwise direction. To connect the line between 0 and '1'. Points 1 and 2 and points 2 and 3 make a counterclockwise angle, but points 3 and 4 make a clockwise angle. So we should not connect points 3 and 4 with a line. Now checkpoints 2 and 4 also make a clockwise angle, so they are not connected. Further, checkpoints 1 and 4 make a clockwise angle so should be connected. Points 4 and 5 and points 5 and 6 make a counterclockwise angle, but points 6 and 7 make a clockwise angle, so points 6 and 7 are not connected, and, further, checkpoint 7–8 and point 8–0 make a counterclockwise angle, and so connect points 4–5, 5–6, and 6–8 with a line. This makes a convex hull.

Jarvis March Algorithm

The steps for the Jarvis March Algorithm can be understood from Figure 3.11:

1. Select the point with the lowest *y*-coordinate as the first coordinate.
2. Select the point with the smallest counterclockwise about the previous vertex.
3. Repeat until reaching the starting vertex.

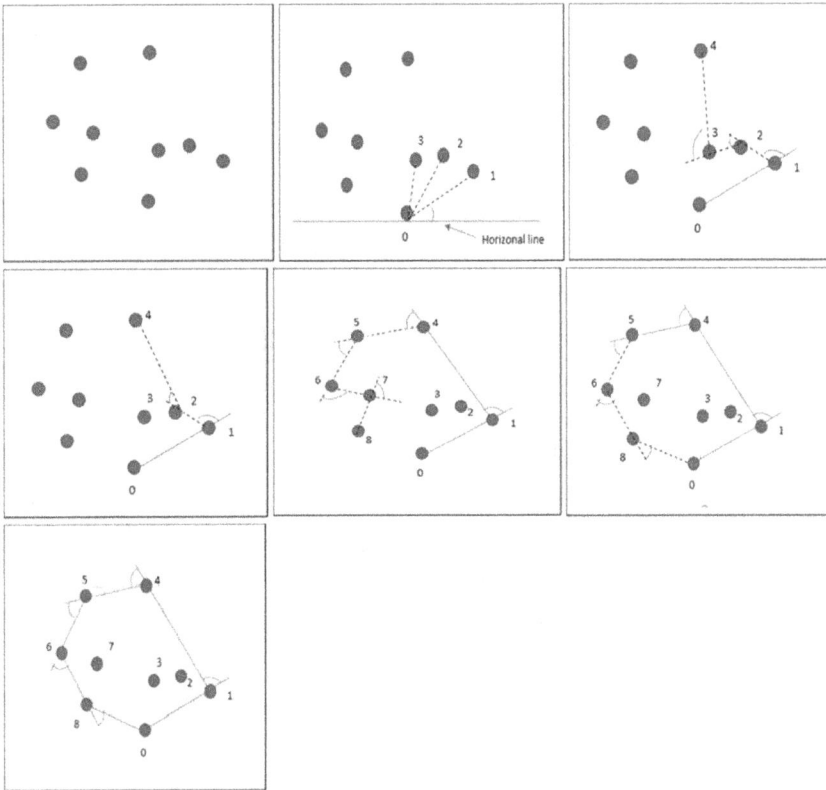

Figure 3.10 Graham scan algorithm.

3.3.3 Support polygon

A support polygon is defined as the polygonal area under all the contact points of feet on the walking surface, or it is the convex hull of the contact point of the foot, as shown in Figure 3.12. For static stability, the center of mass (COM) must pass through a support polygon. In the case of dynamic stability, the center of mass does not always pass through the support polygon.

3.3.4 Zero moment point (ZMP)

In a biped robot, the ZMP is a very important parameter for dynamic stability. ZMP is nothing but the *zero moment point*, the point where all the moment of the forces is zero. For biped robots, during the single support phase (SSP) of walking, it should be under the area of contacting foot. And during the double support phase (DSP), it should be between both feet.

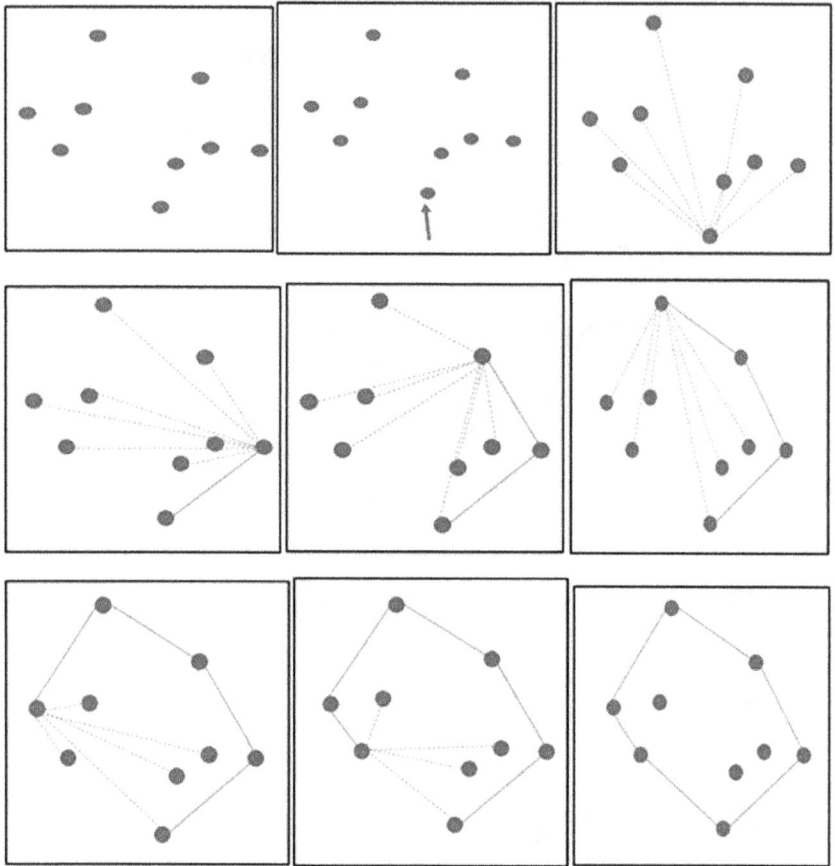

Figure 3.11 Jarvis march algorithms.

Figure 3.12 Support polygon.

During SSP, ZMP is at the center of reaction force. It is in the middle of the foot when the equal distribution of the reaction force is on the foot and at the front part of the foot when the reaction force is dominant in the front part of the foot. During the static and dynamic positions, ZMP should always lie inside the support position. These conditions are shown in Figure 3.13.

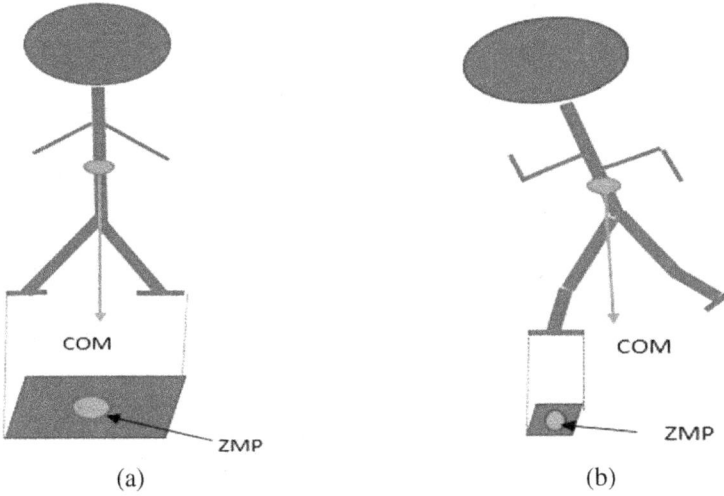

Figure 3.13 (a) ZMP and COM in static condition; (b) ZMP and COM in dynamic condition.

$$x_{zmp}(t) = \frac{\sum_{i=1}^{n} m_i\left(x_i\left(\ddot{z}_i + g\right) - \ddot{x}_i z_i\right)}{\sum_{i=1}^{n} m_i\left(\ddot{z}_i + g\right)}$$

(3.12)

$$y_{zmp}(t) = \frac{\sum_{i=1}^{n} m_i\left(x_i\left(\ddot{z}_i + g\right) - \ddot{y}_i z_i\right)}{\sum_{i=1}^{n} m_i\left(\ddot{z}_i + g\right)}$$

(3.13)

where m_i = mass of ith link, $\ddot{z}_i$ = acceleration of ith link in the z-direction, y_i = acceleration of ith link in the y-direction, x_i = acceleration of ith link in the x-direction, (x_i, y_i, z_i) denotes the coordinate for the center of mass of link i.

3.4 Modeling and simulation

Modeling and simulation are important tools for developing a biped robot. Since the robot has two legs, the walking system makes its control and balance complex compared to the four-legged robot. Developing the physical model before modeling and simulation is costly. It is easy to change the requisite parameters after analyzing the biped model. The present work proposes the biped model simulated on the Simscape Multibody, which is a subset of MATLAB Simulink software. In this simulation model, the 6-degrees-of-freedom biped robot (one DOF ankle, one DOF knee, and one DOF hip joint for each leg) is proposed.

Table 3.2 Biped robot model parameters

	Density	*Shape*	*Size*
Upper leg	1000	Cylindrical	1 cm (radius), 10 cm (length)
Lower Leg	1000	Cylindrical	1 cm (radius) 10 cm (length)
Torso	1000	Cylindrical	1 cm (radius) 10 cm (length)
Foot	1000	Rectangular	4×3×1 (L×B×H)

Modeling is the first step to simulate the virtual model. In this work, the biped model is developed on the Simscape Multibody with the specifications shown in the Table 3.2.

In the model shown in Figure 3.14, the hip, knee, and ankle are connected by the revolute joint with the torso, upper leg, and lower leg, respectively.

3.4.1 Flow chart of modeling and simulation process

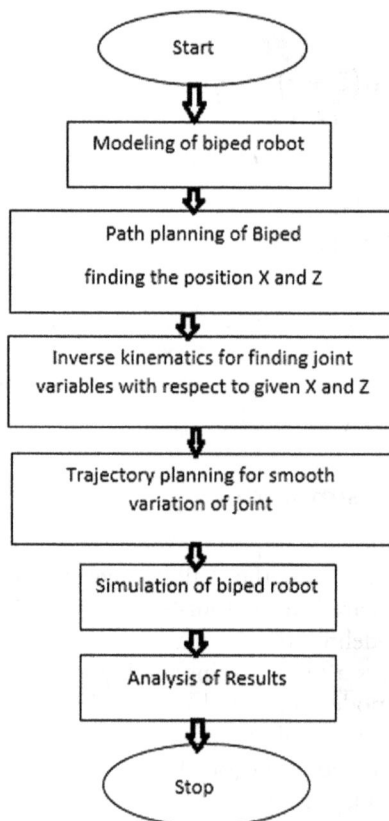

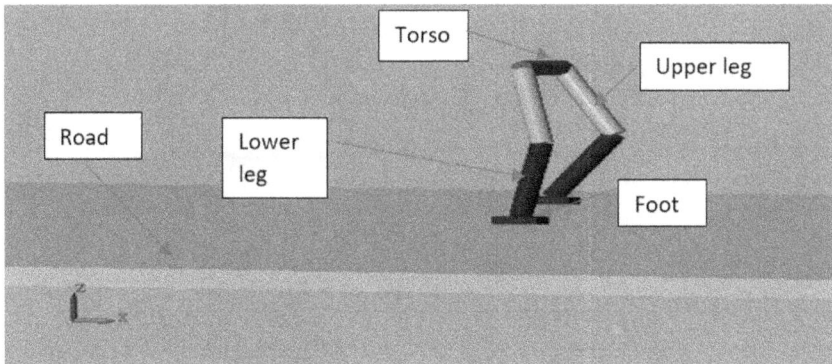

Figure 3.14 Biped robot model.

3.4.2 Modeling process in Simscape Multibody

The basic blocks for making a model in Simscape Multibody are explained next. See Figure 3.15.

Rigid transform
The block is used in the rigid transformation of the coordinate system. The rigid transform block name of plane offset is the coordinate system from which modeling is started.

Solid
The solid command is used to generate solid. In this type of modeling, this command is used to generate the road, foot, lower leg, upper leg, and torso.

6-degree-of-freedom (-DOF) joint
The 6-DOF block is used in this modeling of the movement of the torso in space and is also used for fixing the initial height of the robot from the road before starting the simulation.

Transformation sensor
In this mode, two transformation sensors are used in this model. One is used to find the angular velocity, distance from X, and height of the torso. Stop block is used to stop the simulation when the given condition is satisfied. In this model, the given condition is when the torso height is under 4 cm of the simulation stop.

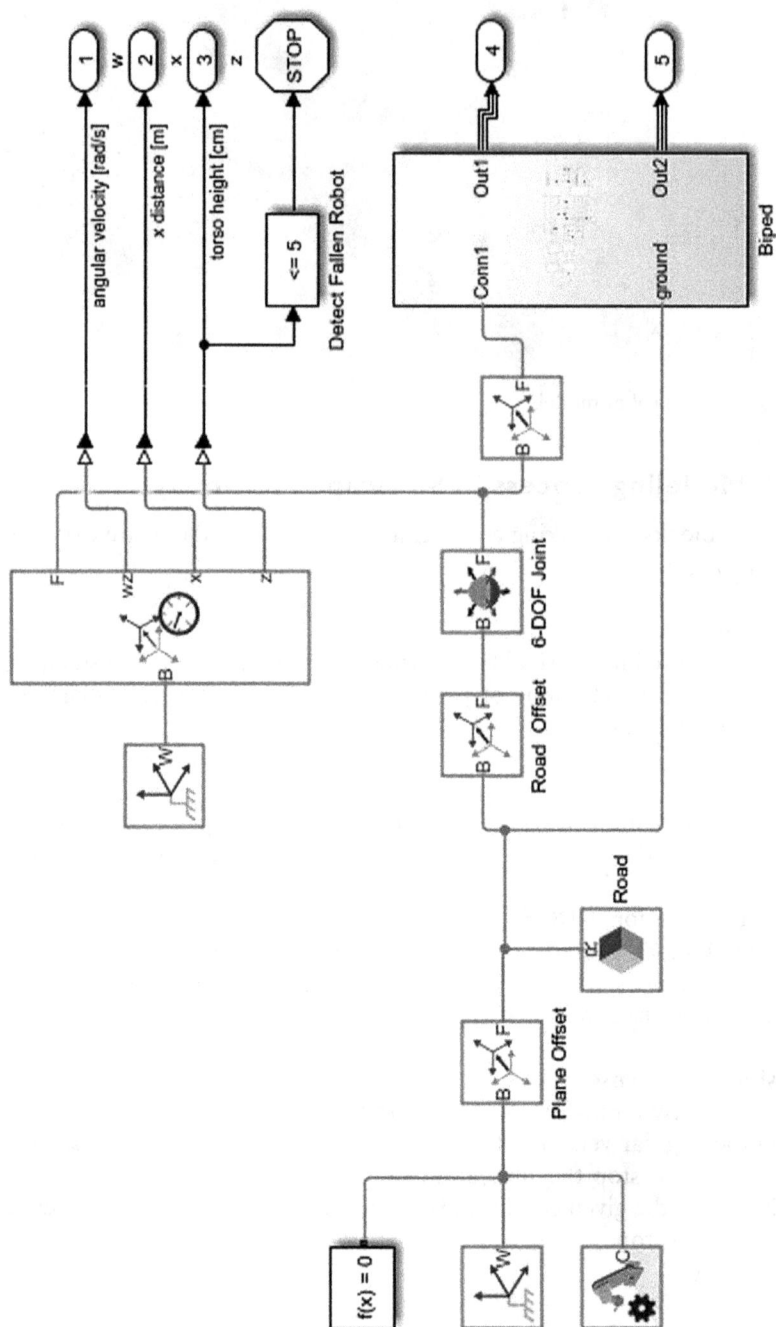

Figure 3.15 Blocks for biped modelling and simulation.

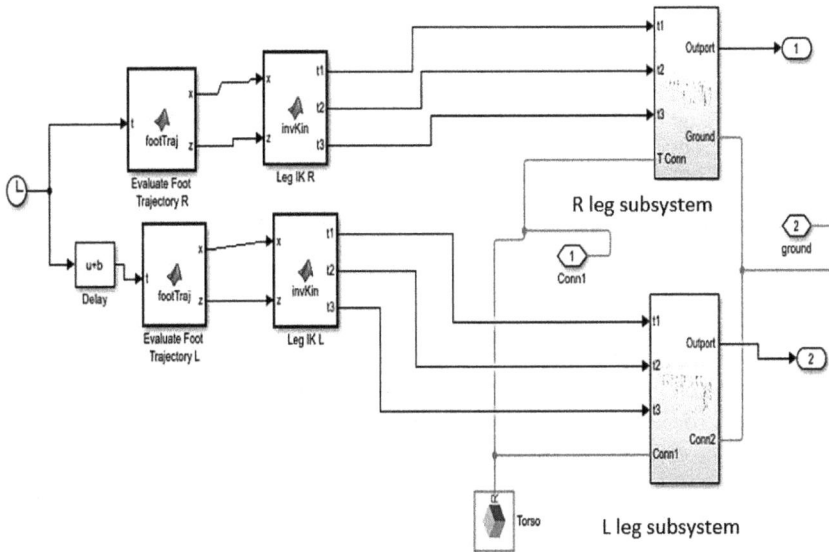

Figure 3.16 Biped subsystem.

Table 3.3 Biped simulation parameters

Parameters	Value
Gait period (*t*)	0.75 sec
Step length (*x*)	0.075 m
Step height (*h*)	0.025 m

Biped robot subsystem

1. *Clock block:* Used for real-time analysis during simulation.
2. *Bias block:* Represented by "Delay" in Figure 3.16. Used to delay the particular block. Here, one leg is delayed by half the gait period time of the other leg.
3. *Function block:* Used to define the function that follows the mathematical equation. In the simulation, the function block is named as "Evaluate Foot Trajectory R," "Evaluate Foot Trajectory L," "Leg IKR," and "Leg IKL," as shown in Figure 3.16.
4. *Simulation parameters:* Table 3.3 shows the parameters used during simulation of the biped robot.

Table 3.4 Contact force parameters

Sphere radius	0.001 m
Road length	25 m
Road height	3 m
Road depth	0.125 m
Contact stiffness	2500 N/m
Contact damping	100
Kinetic coefficient of friction	0.6
Static coefficient of friction	0.8

Contact Force

For walking on the ground, contact force is important. Here, the four spheres at each corner of the foot are used. Using the contact force library, the friction force and normal force are calculated. The parameters used in contact force estimation are shown in Table 3.4.

Revolute joint input and output

The revolute joint is shown in Figure 3.17. The input given to the revolute joint is motion, and angle, speed, and torque are the output. In this model, there are six revolute joints, three in each leg.

In this simulation and modeling of the biped, Simscape Multibody is used. Figures 3.15, 3.16, and 3.17 show the systems and subsystems used for simulation and modeling.

3.4.3 Block diagram of Simscape Multibody

The modeling and simulation of the biped robot block diagram are shown in Figure 3.15. In the figure, the rigid transform block name "plane offset" is used for proper placement of the road. Rigid transform "road offset" is used to offset the position from the road to the 6-DOF joint. The torso is connected to the 6-DOF joint for easy movement in all direction. The 6-DOF joint is also connected with the transformation sensor, which is used for finding the transformation of the torso from the initial position. The torso is inside the biped subsystem that is further connected to the leg subsystem.

Figure 3.16 represents the biped subsystem. It consists of a clock, a delay, a function to evaluate foot trajectory, an inverse kinematic function, torso, left leg subsystem, and right leg subsystem. "Conn1" connects the biped subsystem with the main system presented in Figure 3.16. The torso is connected with conn1, the L leg subsystem, and the R leg subsystem.

The clock block is used to calculate time during the simulation, which is used for determining foot trajectory by means of the evaluated foot trajectory function block, and inverse kinematics is used for calculating joint

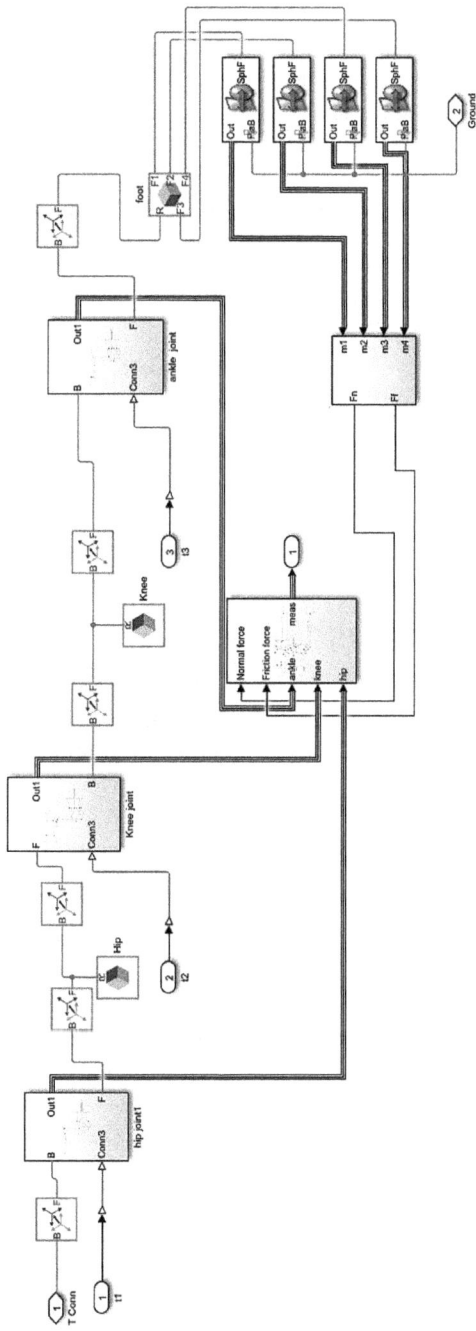

Figure 3.17 Leg sub system.

angles using the Leg "IK R" and "Leg IK L" function blocks. The delay block is used to delay the simulation time. Here the Delay block is used for the delay of one leg with another leg by half of the gait period; in other words, the other leg starts moving after the other leg movement ends.

Figure 3.17 shows the leg subsystem. Two leg subsystems are used in this system: one for the right leg, the other for the left leg. The leg subsystem consists of "U leg," "L leg." Foot. Solid blocks for the model leg and hip joint, knee joint, and ankle joints are revolute joints in this simulation. The "sphere to plane" block is used for finding contact force. Four "sphere to plane" blocks are used at the corner of the foot for calculating contact force; its one end is connected to the foot, and the other end is connected to the ground. The inputs to the hip, knee, ankle joint are $\theta_1, \theta_2, \theta_3$, respectively, which are calculated by using inverse kinematics. The output of the joints are angles, angular velocity, and torque.

3.4.4 Simulation snapshot

The simulation time of a biped robot is 10 sec, and Figure 3.18 shows the image of simulation at a 1 sec gap.

3.5 RESULTS AND DISCUSSIONS

The simulation parameters of the biped robot are as follows: step length (X_f) is 0.075 m, step height (h) is 0.025 m, gait period is 0.75 sec, simulation time 10 sec. In this result, the joint parameter of the right leg of the biped robot is analyzed during walking. Because the movement of the left leg will be similar, only half of a gait period will be delayed so that only the graph of the right leg of the biped is shown. In this simulation of the robot, after the initial position right leg is at stance phase during the first half of gait period; then it is in swing phase.

3.5.1 Joint angles

Figure 3.19 shows the variation of hip, knee, and ankle joint with time. The maximum value of the hip joint is 0.1621 rad, and the minimum value is −5.804 rad. The maximum value comes when the swing leg starts moving, and the minimum value of hip joint comes at the swing half of its step length. These values are repeated after the gait period. The maximum value of the knee joint is 1.082 when the swing leg is at half of the step length. The minimum value is 0.055 rad. This value obtains when the swing leg starts moving.

The ankle joint angle is calculated in such a way that the foot is always in a horizontal position. And it is calculated by $(2\pi - (\Theta_1 + \Theta_2))$, where θ_1 and θ_2 are the hip and knee joint angles, respectively.

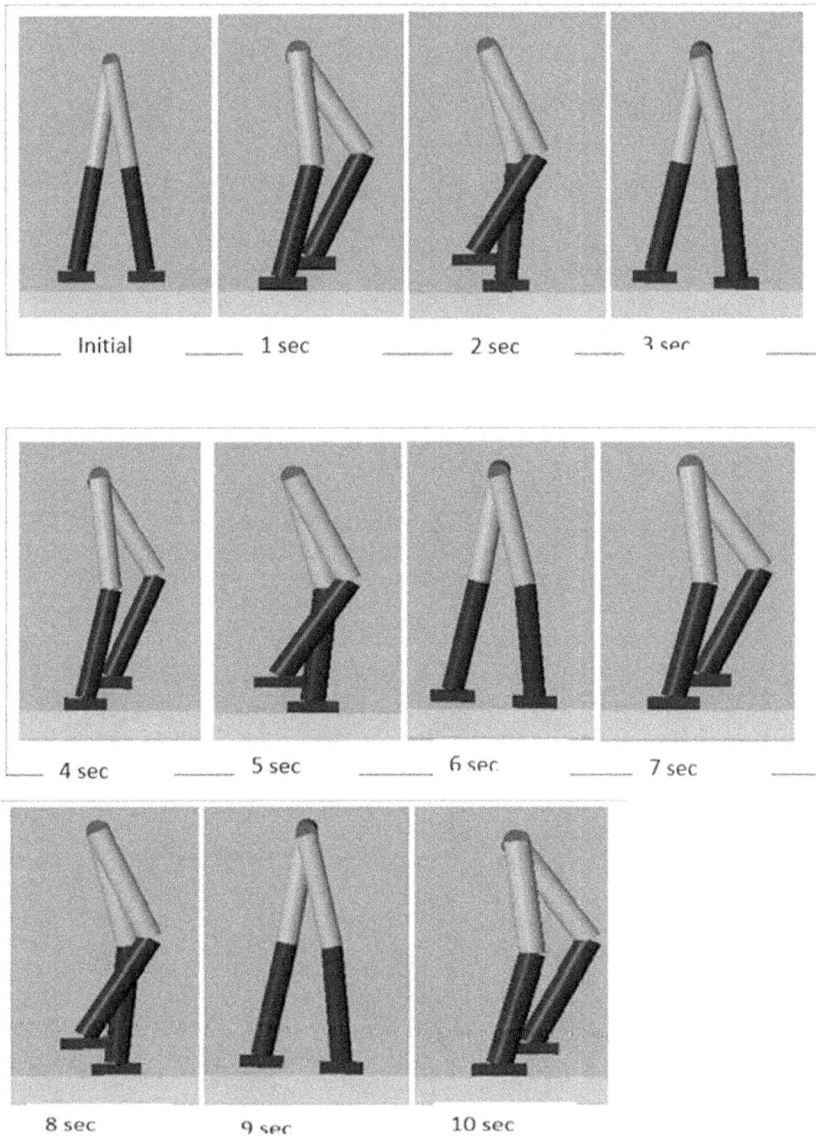

Figure 3.18 Simulation snapshot.

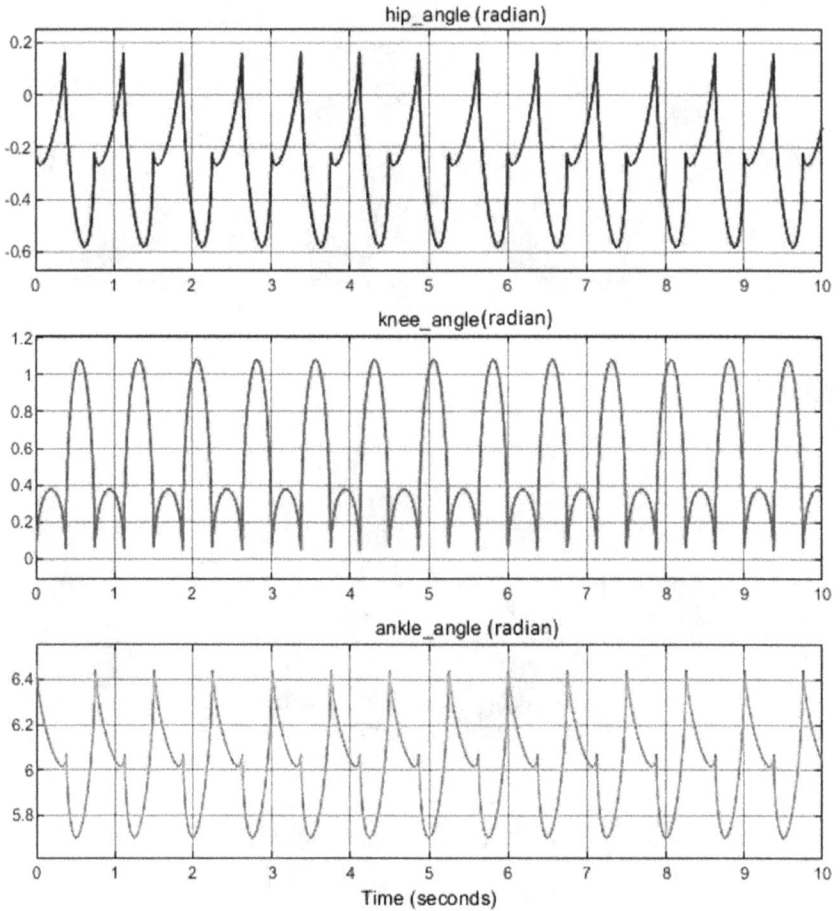

Figure 3.19 Joint angle vs Time.

3.5.2 Joint angular velocity

Figure 3.20 shows the variation of the angular velocity of the hip, knee, and ankle, which is the relative angular velocity of the follower frame concerning the base frame. Ankle angular velocity is a maximum of 26.36 rad/sec with the end of the swing phase of the legs. The minimum ankle speed is –18.41 rad/sec, and it is 0.003 sec after the start of leg swing.

The maximum knee angular velocity is 37.57 rad/sec, and it is 0.003 sec after the start of the leg swing. The minimum knee angular velocity is –51.136 rad/sec at the end of the swing phase. The maximum angular velocity of the hip joint is 25.01 rad/sec when the leg is in a stable position and

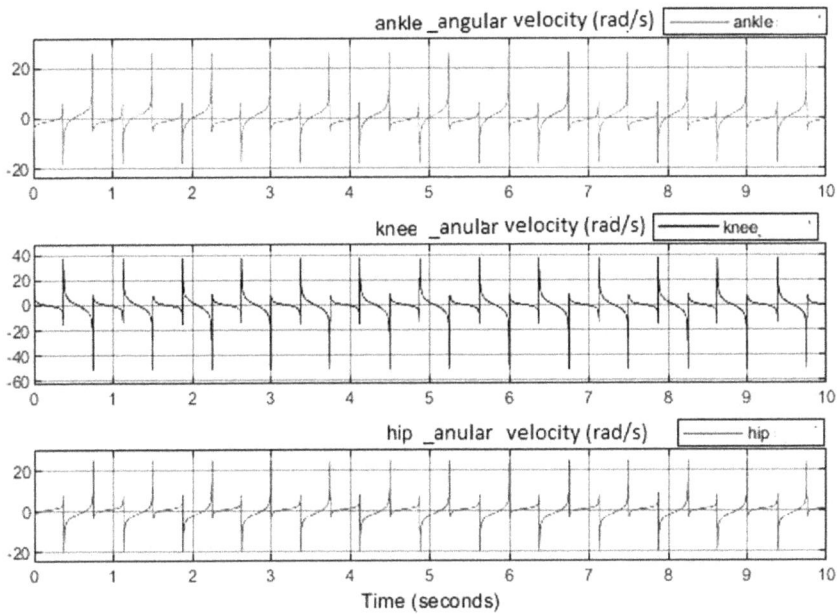

Figure 3.20 Angular velocity vs Time.

the other leg starts to swing. The minimum value is –19.18 rad/sec at 0.003 sec after the leg start swing.

3.5.3 Joint torque

Figure 3.21 shows the variation of the ankle, knee, and hip torque with time. The ankle torque maximum 32.9 N-m at the end of the swing phase. The minimum value of the ankle joint torque (–12.38 N-m) is obtained after 0.003 sec at end of the swing phase and start of the stance phase. The maximum value of the knee joint torque is 37.54 at 0.003 sec after the start of the leg swing. The minimum value of the knee joint torque (–50.83 N-m) at the end of the swing phase and the start of the stance phase. The Maximum value of hip joint torque is 25.90 at the end of the swing leg, or another leg is starting to swing. The minimum value of the hip joint torque is (–19.41 N-m) at 0.003 sec after the leg swing started.

3.5.4 Contact force

Figure 3.22 shows contact force vs. time graph. Using the contact force library of Simulink, the normal force and friction force have been calculated.

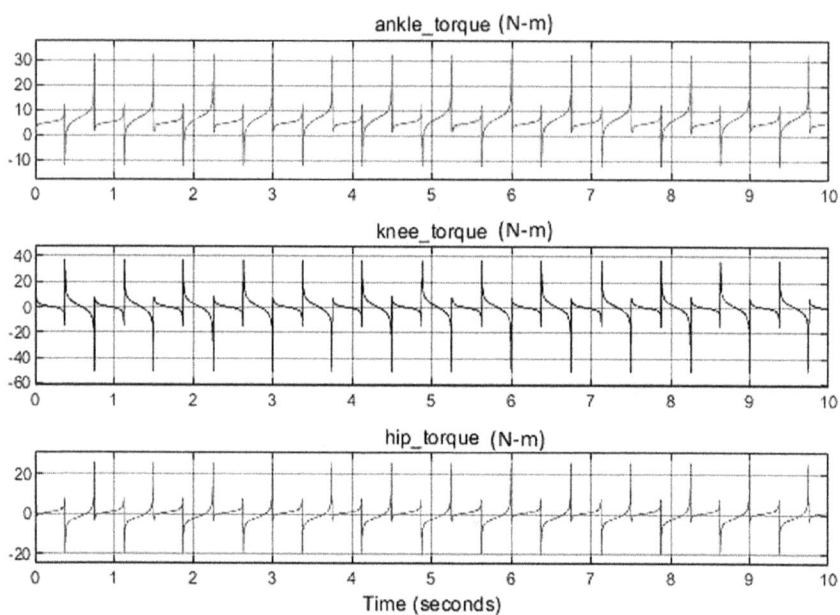

Figure 3.21 Joint torque vs Time.

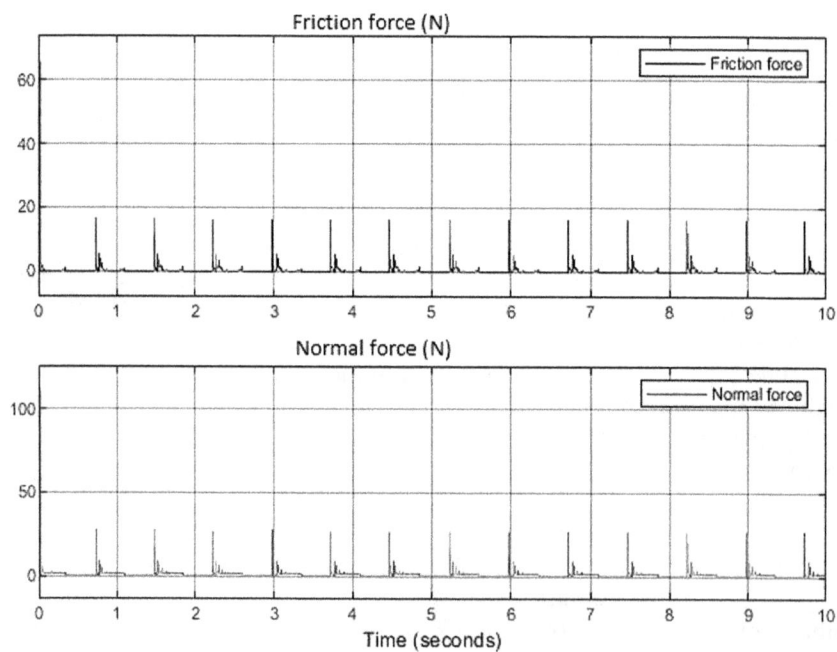

Figure 3.22 Contact force vs Time.

Both forces are maximum at the start because, at the start of the simulation, the initial position of the biped robot is some distance above the ground. When the foot is on the ground at the start of the stance phase, the contact phase is high after the decrease and zero during the swing phase.

3.6 CONCLUSION

The biped robot is modeled and simulated on MATLAB Simulink software. The biped robot maintains its stability during the walking simulation on a flat surface. During walking, the hip joint angle is maximum at the start of the swing phase and minimum when in the middle of the swing phase. The knee angle is minimum at the start of the swing phase. In the middle of the swing phase knee, the joint angle is maximum. The ankle joint angle is minimum at the end of the swing phase. The ankle joint angle is maximum in the middle of the swing phase.

- The angular velocity of the hip joint is maximum at end of the swing phase or at the start of the stance phase. The hip joint angular velocity is minimum at 0.003 sec after the start of the swing phase. The knee joint angular velocity is minimum at the end of the swing phase. The knee joint angular velocity is maximum at 0.003 sec after the start of the swing phase. The ankle joint angular velocity is maximum at the end of the swing phase and minimum at the start of the swing phase.
- The hip joint torque is maximum at the end of the swing phase. The hip joint torque is minimum at 0.003 sec at the start of the swing phase. The knee joint torque is minimum at the end of the swing phase
- The knee joint torque is maximum at 0.003 sec at the start of the swing phase. The ankle joint torque is maximum at the end of the swing phase and minimum at 0.003 sec at the start of the swing phase.
- Contact forces are maximum at the start of simulation because at the start of the simulation, the initial position of the biped robot is some distance above the ground. When the foot is on the ground at the start of the stance phase, the contact phase is high after the decrease and zero during the swing phase.

3.6.1 Future scope

- The method implemented in research work for simulation of the biped is applicable for real experimental tests.
- The feedback control system should be in place so that the accurate path can be followed as presented.
- Significant efforts can be made for optimization of the path with an additional objective of the lowest energy consumption during walking.

- In order to mimic real humanoid motion, efforts can be made to increase the degree of freedom in the leg joint, so that the robot can make smoother and more stable motion.
- Along with the lower body, there should be upper body motion for future scope so that the robot moves like a human. Significant work has to be done in this area.

REFERENCES

[1] G. Gupta and A. Dutta (2018), "Trajectory generation and step planning of a 12 DoF biped robot on uneven surface," *Robotica*, vol. 36, no. 7, pp. 945–970, DOI: 10.1017/S0263574718000188.

[2] A. Ankarali (2012), "Fuzzy logic velocity control of a biped robot locomotion and simulation," *Int. J. Adv. Robot. Syst.*, vol. 9, DOI: 10.5772/52555.

[3] Abhishek Sarkar and Ashish Dutta (2019), "Optimal trajectory generation and design of a compliant biped robot for walk on inclined ground," *Journal of Intelligent & Robotic System*s, 94,583–602, doi:10.1007/s10846-018-0882-9

[4] M. M. Moghaddam, A. Hadi, A. Tohidi, and M. Elahinia (2011), "Design, modeling, and prototyping of a simple semi-modular biped actuated by shape memory alloys," *J. Intell. Mater. Syst. Struct.*, vol. 22, no. 13, pp. 1489–1499, doi: 10.1177/1045389X11418861.

[5] P. R. Vundavilli and D. K. Pratihar (2011), "Near-optimal gait generations of a two-legged robot on rough terrains using soft computing," *Robot. Comput. Integr. Manuf.*, vol. 27, no. 3, pp. 521–530, doi: 10.1016/j.rcim.2010.09.007.

[6] J. C. Lu, J. Y. Chen, and P. C. Lin (2013), "Turning in a bipedal robot," *J. Bionic Eng.*, vol. 10, no. 3, pp. 292–304, doi: 10.1016/S1672-6529(13)60225-5.

[7] C. Sun, W. He, W. Ge, and C. Chang (2017), "Adaptive Neural Network Control of Biped Robots," *IEEE Trans. Syst. Man, Cybern. Syst.*, vol. 47, no. 2, pp. 315–326, doi: 10.1109/TSMC.2016.2557223.

[8] P. N. Mousavi and A. Bagheri (2007), "Mathematical simulation of a seven link biped robot on various surfaces and ZMP considerations," *Appl. Math. Model.*, vol. 31, no. 1, pp. 18–37, doi: 10.1016/j.apm.2006.06.018.

[9] R. Panwar and N. Sukavanam, 2020, "Trajectory tracking using artificial neural network for stable human-like gait with upper body motion," *Neural Comput. Appl.*, vol. 32, no. 7, pp. 2601–2619, DOI: 10.1007/s00521-018-3842-1.

[10] R. K. Mandava and Vundavilli (2018), "Implementation of modified chaotic invasive weed optimization algorithm for optimizing the PID controller of the biped robot." Sadhana 43,66, DOI:10.1007/s12046-018-0851-9.

[11] Y. Sugahara *et al.*, 2005, "Walking up and down stairs carrying a human by a biped locomotor with parallel mechanism," *2005 IEEE/RSJ Int. Conf. Intell. Robot. Syst. IROS*, no. 16, pp. 3425–3430, DOI: 10.1109/IROS.2005.1545500.

[12] X. Zang, Z. Lin, Y. Liu, X. Sun, and J. Zhao, 2017, "Control strategy research for a biped walking robot with flexible ankle joints," *Proc. – 2017*

1st IEEE Int. Conf. Robot. Comput. IRC 2017, no. grant 51675116, pp. 93–96, DOI: 10.1109/IRC.2017.9.

[13] MathWorks Student Competitions Team (2022). MATLAB and Simulink Robotics Arena: Walking Robot (https://github.com/mathworks-robotics/msra-walking-robot), GitHub.

[14] Prasad, R. and Arif, M., 2018. "Workspace and Singularity Analysis of Five Bar Planar Parallel Manipulator". *2018 5th IEEE Uttar Pradesh Section International Conference on Electrical, Electronics and Computer Engineering (UPCON).*

Chapter 4

Design and analysis of heavy vehicle medium duty transmission gearbox system

Ashwani Kumar, Yatika Gori, Chandan Swaroop Meena, Varun Pratap Singh, and Vipin Kumar Sharma

4.1 INTRODUCTION

In the growing economy of India, the manufacturing sector has a very important role. Goods, food products, construction materials, automobile components, etc. are moved from manufacturing place to service place, where it is distributed in society. For transporting these products and services, there is the requirement of the transportation medium. Heavy vehicles have been used for a long time for transportation purposes, and these vehicles travel over thousands of kilometers. In heavy vehicles, medium duty heavy vehicles are most suited for service within cities. Therefore, in considering the importance of these vehicles, it is mandatory that these vehicles are always in good operating condition. To eliminate the chances of failure of these vehicles, a thorough examination of the vehicle is required. A new transmission system has been designed for heavy vehicle medium duty trucks, and it is investigated thoroughly.

The automobile transmission system is a combination of gears to meet the torque variations for different speed conditions. The transmission system can be classified in three types: automatic, manual, and continuously variable

DOI: 10.1201/9781003360001-4

transmission. The simplest type of transmission is manual. Manual transmission is of two types: sliding mesh and constant mesh. In automatic and continuously variable transmissions, there is less driver interaction. Slack in the drive train mechanism produces high vibration, known as transmission shock.

The transmission system is the most important part of the vehicle assembly. The truck transmission casing, or housing, is subjected to vibration induced by the load fluctuations, gear defects, harmonic excitation, meshing excitation, and varying speed and torque conditions. The vibration of the casing radiates noise. Noise and vibration are the two technical indexes for transmission failure. This casing vibration is the result of internal excitation caused by the gear meshing process. This internal excitation produces a dynamic mesh force, which is transmitted to the casing through the shafts and bearings. For the accurate prediction of dynamic behavior, the value of the mesh force transmitted to the casing should be considered, but it is difficult to obtain the value of mesh forces by the experimental or numerical method. The objective of this chapter is analysis of the vibration signature for different boundary-conditions-based problems. For this purpose, the numerical simulation method finite element analysis (FEA) has been selected. This numerical method allows us to compute the generalized mesh forces transmitted to the casing and to determine the main responsible force for the vibrations of the casing. The comparison between the results obtained by the numerical simulation shows the effectiveness of the work. It is required to find the natural frequency and mode shape for the accurate prediction of transmission housing life and to prevent it from damage. Transmission housing is made of material such as cast iron, Al alloys, Mg alloys, structural steel, and composites. Noise and vibration reduction in the heavy vehicle transmission system is a constant development goal in automotive engineering. The vehicle transmission system determines the overall level of noise, together with the engine, bodywork, and chassis. The rattling and clattering noise are caused by vehicle transmissions under torsional vibration conditions and are produced by loosened parts vibration of transmission components moving backward and forward within their clearances when not under load. The reciprocity principle is used to determine the failure frequencies for the transmission casing.

A critical literature review was performed to study the previous research. Dogan [1] has investigated the cause of rattling and clattering noise and concluded that torsional vibration is the main reason for the vibration. For this analysis, Dogan used simple gearbox geometry consisting of only the transmission casing. The main advantage of the Dogan study is that he started simulating such a complex geometry of the transmission gearbox. He used the EKM simulating program. Wang et al. [2] studied the excitation phenomena of the gearbox. Abouel-Seoud et al. [3] performed a similar study on the car gearbox. They used a vibration response analysis method for the analytical analysis of the car gearbox system. They performed analytical and experimental analysis of a car transmission system. By using physical properties, they calculated the radiation efficiency, and the vibration response was measured. Vandi et al. [4] presented the implementation of a simplified engine-driveline model to complete an existing vehicle dynamic model. The engine model is based on maps that are expressed as a function of engine speed and load.

Nacib et al. [5] performed the failure analysis of the heavy gearbox of helicopters. To prevent breakdowns and accidents in helicopter, gear fault detection is important. Spectrum analysis and Cepstrum analysis were used to identify damaged gear. Gordon et al. [6] studied the source of vibration. A sports utility vehicle with a sensor and data acquisition system is used to find the vibration source. This study was focused on vehicle vibration response from road surface features. Kar et al. [7] used motor current signature analysis (MCSA) and discrete wavelet transform (DWT) for studying gear vibration. Transmission errors and internal excitation cause vibration and noise problems. Czech [8] described the vibroacoustic diagnostics of high-power toothed gears. The presented analysis is an experimental work done in a steel plant. Singh [9] did two case studies for the vibroacoustic analysis of automotive structures. Analytical and experimental results are presented for a brief description. In the first case, passive and adaptive hydraulic engine mounts and in the second case, welded joints and adhesives in vehicle bodies were considered. Tuma [10] investigated the noise and vibration problems in TATA trucks. The Fourier transform was used for analytical study, and the experimental results find the heavy vibration frequency zone (500–3000 Hz). Kumar et al. [11] studied the modern transmission system.

Yu et al. [12] have studied the dynamic characteristic of the simple transmission gearbox casing with constraint bolt position. Gray Cast Iron HT200 was used as the transmission casing material. The finite-element-method- (FEM-) based simulation method was used, and the simulation result was verified with experimental results. For experimental analysis, the transmission casing was constrained on a hanging base. The excitation was provided using a hammer. This research work was performed on the simple geometry of the transmission gearbox housing. This work can be extended to analysis of the full transmission gearbox assembly. The full transmission system consists of the gearbox and housing, and finite element analysis (FEA) simulation is performed on these components. Yu et al. [12] studied the transmission gearbox housing using only one material, Gray Cast Iron HT200. The possible use of other materials like Gray Cast Iron FG 260, Al alloys, and steel alloys AISI 4130 can be explored. To continue this study, the modern transmission system of a heavy vehicle was studied for looseness analysis, structural analysis, and variation of housing material [13–15].

Kostić et al. [16] investigated the natural vibrations of the housing walls and concluded that it can be prevented by design parameters. The heavy vehicle medium duty truck transmission system was selected as the research objective for this chapter in place of TATA trucks. Modal and thermal characteristics analysis were studied for heavy vehicle medium duty trucks [17–18]. The dynamic response was studied for different objects using ANFIS [19–22]. This study was useful in conducting the dynamic vibration study of the heavy vehicle medium duty transmission gearbox [26–29]. Abbes et al. [30] conducted the dynamic analysis of gearbox housing. Transmission error was studied with the help of FEA. The natural frequency of gearbox housing varies (285–2210 Hz). A stretcher was introduced in the gearbox housing to control the vibration. The lower frequency causes a resonance condition leads to failure of the system.

The literature study shows that authors have used the response surface method (RSM) in vehicle parts analysis, vibroacoustic analysis, damping structure analysis, welding parameters optimization, rapid prototyping parameters optimization, etc. Vibroacoustic design was optimized using RSM [31–35]. The sound level inside the passenger cabin was studied using a FEM-BEM combination. For structural analysis, FEM was used. The boundary element method (BEM) was coupled with FEM for acoustic analysis. Vehicle modeling and optimization were studied [36–38]. The authors studied vehicle concept modeling using the optimization approach to improve the noise and vibration, stiffness, crash performance, etc. Morphing was used to study the vehicle structure [39–41]. Kumar et al. [42] used the FEA technique for single-piece drive shaft analysis. They measured the performance of the drive shaft by varying load and torque conditions. The drive shaft is an important component of the power train system.

4.2 CLASSIFICATION OF TRUCKS

Trucks are available in different styles and are divided into many types depending on the trucking authorities. Some classifications are accepted globally. One of the globally accepted classifications is based on gross vehicle weight rating (GVWR). Using GVWR classification, trucks are classified mainly in three types: light duty trucks, medium duty trucks, and heavy duty trucks (Table 4.1).

Light duty trucks are those having the lightest hauling capacities of goods and cargo. The regular cargo transporters use these vehicles, which are the best examples of vehicles having better traveling abilities. Light duty trucks have a GVWR of 0–4536 kg (0–10,000 lb). Different manufacturers, Titan, Nissan, GMC, and Ford, manufacture these vehicles. Some of the most popular models of this category are Tacoma, Sonoma, Titan, and F250. SUVs and tow, salvage, flatbed, and dump trucks are popular types of models.

Table 4.1 Classification of trucks

Type	Gross vehicle weight rating (GVWR)
Light duty trucks	0–4536 kg
Medium duty trucks	**4536–8845 kg**
Heavy duty trucks	8845–14968 kg

Medium duty trucks have aggressive hauling capacities of 4536–8845 kg (10,001–19,500lb). Medium duty vehicles are popular for transportation of raw materials, passing of industrial semifinished or finished products, and getting goods to the ultimate users. Medium duty trucks are the most frequently used and demanded trucking types by transporters. These trucks have different specialties like dual rear wheels or even four rear wheels. The prime manufacturers are Ford, GMC, Tata Motors, and Mahindra. Crane, garbage, lift, hook lift, and pickup trucks are some of the most popular types of medium duty vehicles.

Heavy duty trucks have highest hauling power of trucks and are considered specifically well suited for carrying the heaviest cargo loads. These vehicles have capacities of 8845–14,968 kg (19,501–33,000 lb). The international manufacturers of these trucks is International, Mack, Kenworth, and Peterbilt. Beverage trucks, chipper trucks, fire trucks, packer trucks, semitrucks, and big trucks are some of the types in the heavy duty vehicles category.

The prime manufacturers of medium duty trucks in India are Tata Motors and Mahindra. Table 4.2 shows the power and torque ranges with clutch, gearbox, and GVWR specifications of medium duty trucks. This study is used to find the power and torque ranges for simulation. From Table 4.2, the power range for medium duty trucks is 51.9–91.94 kW and the torque range is 200–400 Nm. This range of power and torque will be used in the FEA simulation. In this chapter, a broad range of power and torque is covered to obtain accurate and useful results at high and low rpm.

4.3 DESIGN SPECIFICATION OF TRANSMISSION GEARBOX SYSTEM

From the survey of Ashok Leyland, Tata Motors and Mahindra heavy vehicles, the following specifications were identified for the current gearbox model. Figures 4.1–4.4 show the different stages in development of the transmission gearbox system.

Table 4.2 Power and torque range for medium duty trucks

Truck	Power	Torque	Clutch	Gearbox	GVWR
TATA: LPK 407 Tipper	100 PS (73.54 kw) @ 2800 rpm	300 Nm @ 1400–1500 rpm	Single-plate dry friction type (280 mm)	GBS-27 Synchromesh, 5F + 1R	6250 kg
TATA: SFC 407 EX TT	75 PS (55.16 kw) @ 3050 rpm	245 Nm at 1400–1600 rpm	Single-plate dry friction type (280 mm)	G-380 Synchromesh, 5F+1R	5300 kg
TATA: SFC 407 EX HT	100 PS (73.54 kw) @ 2800 rpm	300 Nm @ 1400–1500 rpm	Single-plate dry friction type, 280 mm (Reinforced)	GBS-27 Synchromesh, 5F+1R	6250 kg
TATA: SFC 709 EX	125 PS (91.94 kw) @ 2400 rpm	400Nm @ 1300–1500 rpm	Single-plate dry friction type (310 mm)	GBS-40 Synchromesh	7490 kg
TATA: LPT 407	100PS (73.54 kw) @ 2800 rpm	300 Nm at 1400–1500 rpm	Single-plate dry friction type, 280 mm	GBS-27 Synchromesh, 5F+1R	7250 kg
TATA: LPT 709 EX	125 PS (91.94 kw) @ 2400 rpm	400 Nm @ 1300–1500 rpm	Single-plate dry friction type	GBS40 Synchromesh 5F+1R	7490 kg
TATA: LPT 709 HEX2	125 PS (91.94 kw) @ 2400 rpm	400 Nm @ 1300–1600 rpm	Single-plate dry (310 mm)	GBS 40 Synchromesh, 5F + 1R	8720 kg
TATA: SK 407 EX	100 PS (73.54 kw) @ 2800 rpm	300 Nm @ 1400–1500 rpm	Single-plate dry friction type, 280 mm (reinforced)	GBS-27 Synchromesh, 5F + 1R	6250 kg
Mahindra DI 3200 CRX	51.9 KW (70 HP) @ 3200 rpm	200 Nm@ 1800–2200rpm	Single-plate dry friction type	5F+ 1R With Over Drive	4600 kg
Mahindra Load King Zoom (4 & 6 tire)	58.5 KW (78 HP) @ 3200 rpm	205 Nm @ 1800–2200 rpm	Single-plate dry friction type, 240 mm	5F + 1R	6255 (4 tire)/ 5950 (6 tire)
Mahindra Load King Zoom Tipper	58.5 KW (78 HP) @ 3200 rpm	205 Nm @ 1800–2200 rpm	Single-plate dry friction type, 240 mm	5F + 1R	5950 kg

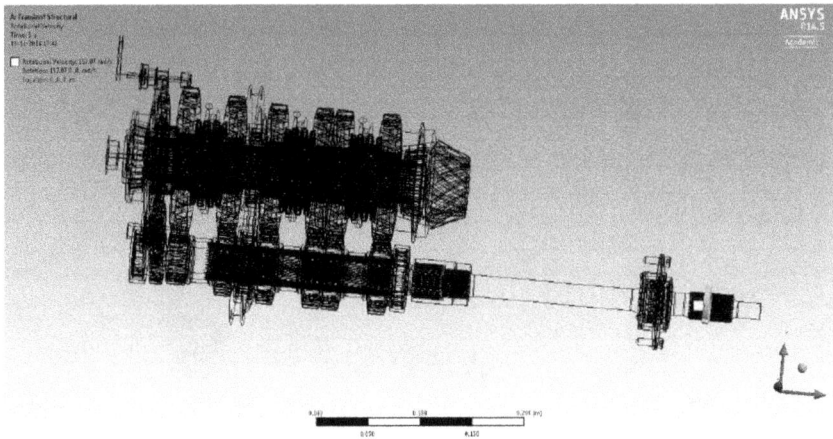

Figure 4.1 Wire frame model of gearbox assembly.

- Heavy vehicle transmission gearbox type: Synchromesh, 4 forward + 1 reverse
- Engine to which transmission assembly can be fitted: Turbo intercooled
- Maximum power output: 55.2 kW @ 3050 rpm
- Maximum torque generation: 245 Nm (1400–1600 rpm)
- Maximum speed in top gear: 110 kilometer per hour (kmph)
- RPM variation: 1000–7000
- Temperature of transmission housing: Atmospheric temperature (20–50°C)
- Internal temperature of transmission housing: 70–120°C
- Clutch gearbox assembly: Single-plate dry friction type (280 mm)
- Steering: Mechanical steering
- Steering ratio: 25–28
- Wheelbase: 3100 mm (high deck).
- Front axle: Heavy duty forged I beam, reverse Elliot type
- Rear axle: Single reduction, hypoid gears, fully floating axle shafts
- Fuel injection pump: Rotary type (mechanically governed)
- Gross vehicle weight (GVW): 5.95 T (medium duty).

4.3.1 Modeling and analysis

Solid Edge and Pro-E software [24, 25] was used for the 3D solid modeling of transmission gearbox components. The transmission gearbox assembly was designed in two major steps. In the first step, the gearbox was designed, and in second step, the gearbox system was fitted with the housing. The modeling of a heavy vehicle medium duty truck transmission gearbox consists of

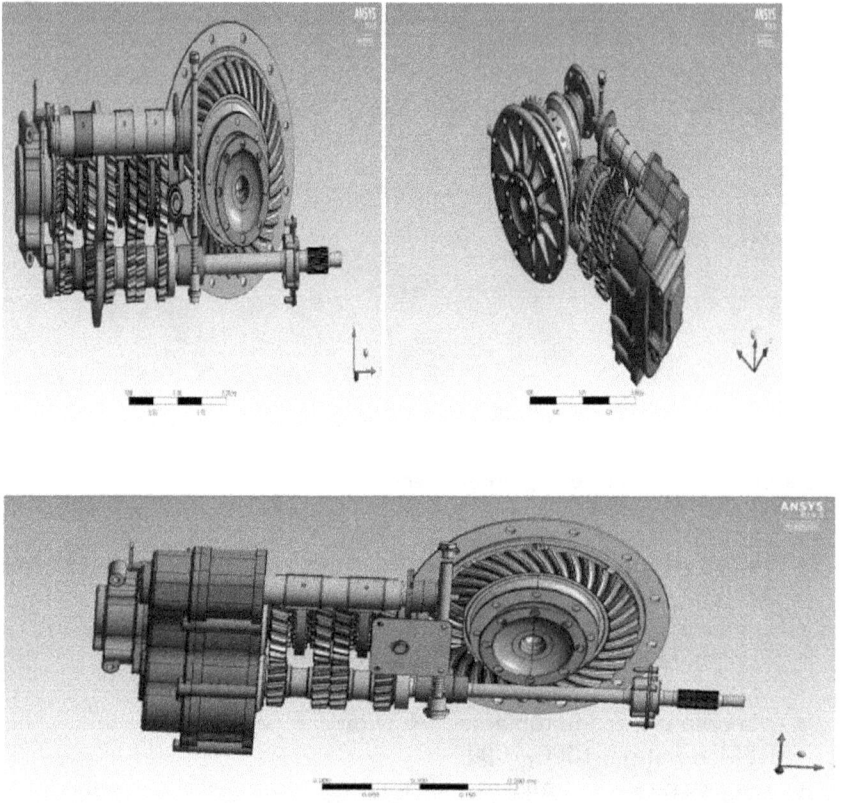

Figure 4.2 Transmission gearbox components.

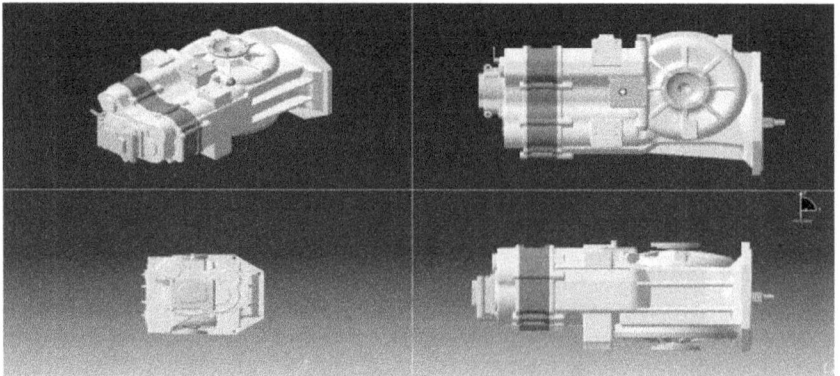

Figure 4.3 Different views of transmission gearbox system.

Figure 4.4 Full transmission gearbox assembly of heavy vehicle medium duty truck.

more than 600 parts. The assembly of the transmission gearbox consists of shafts, gears and mountings, etc. In the present model, various reinforcing ribs, drain hole, corners, bosses and fillets, and connecting bolt holes were considered. In order to get the analysis results close enough to the actual running conditions, only a few parts of the gearbox were ignored in order to simplify the design; these parts have no effect on natural frequency.

Ansys R 14.5 [23], single user with unlimited node facility, was used as an analysis tool. Ansys 14.5 is based on FEA (finite element analysis), which works on the nodes and elements concept. Elements are connected at a point known as a node. This process is known as meshing. The meshed model of a transmission housing consists of 196,137 nodes and 113,566 elements. A linear tetrahedron element (Tet 4) was used for meshing. The mesh model is generated during FEA simulation. It discretizes the analysis object in small parts. These are known as elements, and elements are connected to one another using nodes. The FEA meshed model of the Gray Cast Iron HT200 housing consists of 196,137 nodes and 113,566 elements. The gear assembly is meshed using 575,383 nodes and 339,898 elements. A single-piece drive shaft is meshed using nodes (87,718) and elements (453477). FEA-based Ansys was selected as an analysis tool because it has broad application in the engineering field. It is used for structure, modal, thermal, transient, electromagnetic, and fluid–solid interaction problems. The results obtained from Ansys are reliable and can be used a field application. In literature, Abbes et al. [30] investigated the effect of transmission error on the dynamic behavior of a gearbox housing using the DSM (digital surface model) and FEM methods, and they concluded that the FEM method provides more accurate results.

Design Expert, Version 9.0, was used for response surface optimization due to its broad application in DSM engineering field and accurate results.

4.4 DYNAMIC PERFORMANCE EVALUATION OF TRANSMISSION GEARBOX

The main objective of this chapter is to investigate the performance of a vehicle's transmission gearbox under the influence of load, rotational speed, and lubrication on a multispeed gearbox gear surface. Gear oil SAE 80W-90 was used as the gearbox lubricant, for the cooling of the transmission gearbox for high performance. An assumption has been made that the air–gear oil mist within the transmission is under a steady-state condition, in isothermal equilibrium with the transmission gear oil bath of the lubricant. The lubrication in the multispeed transmission is subjected to thermoelastohydrodynamic lubrication. The present chapter deals with the thermomechanical performance study of a multispeed transmission (4 speeds, excluding reverse gear) system that combines the transient structure analysis of the gear train assembly.

The overall vehicle transmission gearbox performance is governed by the gear oil properties, and it also effects fuel consumption. Transmission gear oil viscosity highly depends on temperature, it varies exponentially, and the other properties vary linearly. This research study was performed at high loading, 245 Nm, 1500 rpm rotational speed, and (100–600) W/m²k convection heat transfer variation. At each loading condition, the thermal profile of the transmission gearbox surface was evaluated using steady-state thermal analysis in isothermal equilibrium. The analysis result shows that the gearbox oil thermal properties directly affect the performance and life span of the automobile transmission gearbox.

In this research work, efforts were made to measure the surface temperature variation of the gearbox surface due to gear oil temperature, frictional conditions, and dynamic loading under the influence of SAE 80W-90 lubricating gear oil. The properties of SAE 80W-90 are: density 887 kg/m³ (15.6°C); viscosity 139cSt (40°C), 15 cSt (100°C); viscosity index 110; flash point 218°C; pour point (–27°C).

4.4.1 Case 1: Gear oil bath temperature (80°C) and average convection heat transfer (h = 100–600 W/m²k)

Finite element analysis was used for the structural and thermal simulation of the transmission gearbox. The internal temperature of the gearbox has an influence on thermal stresses, and deformations lead to failure. Figure 4.5 shows the temperature variations at different points of the gearbox surface for gear oil bath temperature of 80°C. Figure 4.5(a) highlights the temperature variation in the gear train assembly at the value of $h_{100 \text{ W/m}^2 \text{ k}}$ for convective heat transfer coefficient (h). The temperature variations are shown in color code. The minimum temperature is 342.94 K, and the

(a) Temperature variation at h = 300 W/m^{2}k

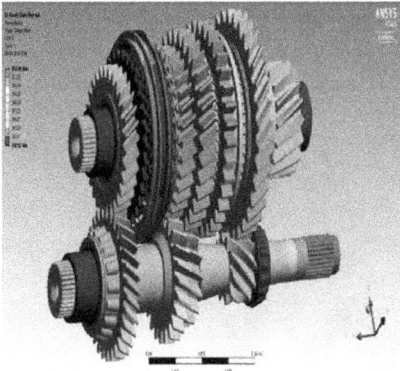

(b) Temperature variation at h = 400 W/m^{2}k   (c) Temperature variation at h = 500 W/m^{2}k

(d) Temperature variation at h = 600 W/m^{2}k

Figure 4.5 Gearbox surface temperature variation: isothermal gear oil bath temperature (80°C).

(a) Temperature variation at $h = 200$ W/m^2 k

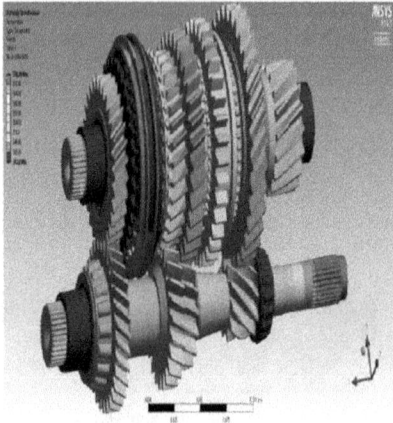

(b) Temperature variation at $h = 300$ W/m^2 k

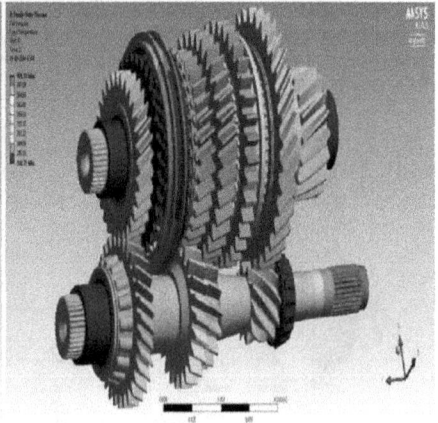

(c) Temperature variation at $h = 400$ W/m^2 k

(d) Temperature variation at $h = 500$ W/m^2 k

Figure 4.6 Temperature profile variation on gearbox surface: isothermal gear oil bath temperature (100°C).

maximum is 350.01 K. The maximum temperature effect is found on the countershaft's right end in red hues, which was fixed in simulation. Red hues signify the maximum temperature change and thermal stress generation portion. Between minimum and maximum temperatures, eight other temperatures are shown in the figure. The temperatures vary very gradually at different points, as can be seen from Figure 4.7. In Figure 4.7, h_{100} (1) shows the gradual increment in temperature at different points of the gear train. Around the second gear pair, the temperature and deformation are high. Figure 4.5(b) shows the temperature variation in the gear train assembly at the value of $h_{200\ \text{W/m}^2\ \text{k}}$. As the value of h increases, there is very small change in the temperature profile of the gear train. The minimum temperature decreases 0.09 K, and the maximum temperature increases by 1.58 K. The temperature profile varies between 342.85 and 351.59 K. The high temperature region is at the same place of the fixed portion at the right side end of the counter shaft. Figure 4.5(c) shows the temperature increment on the third gear. The temperature reached 350.01 K. On the second gear, the temperature is 349 K under the influence of $h_{300\ \text{W/m}^2\text{k}}$. The temperature profile varies between 342.93 352.03 K. Figure 4.5(d) shows the gear train temperature profile varies 342.92 to 352.38 K. The first gear is in loose meshing, and its temperature is 347.12 K. The second and third gears show a temperature increment and have a value of 349.23–351.33 K. The lay/countershaft also shows a variation of temperature (346.07–352.38 K) for $h_{400\ \text{W/m}^2\text{k}}$. The third gear profile shows a temperature increment in red hues,

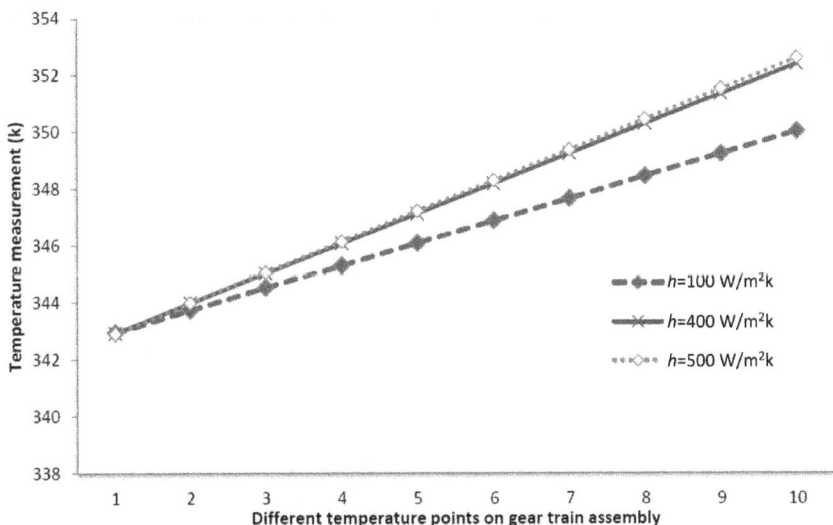

Figure 4.7 Temperature variations on gearbox surface at oil bath temperature 80°C with varying *h*.

and this area is prone to thermal stresses more for $h_{500\ \text{W/m}^2\text{k}}$, $h_{600\ \text{W/m}^2\text{k}}$ convective heat transfer coefficient values. In Figure 4.7, h_{400} and h_{500} show the gear train temperature profile variations.

4.4.2 Gear oil bath temperature (100°C) and average convection heat transfer (h = 100–600) W/m²k

For the second part of study, the gear oil temperature was increased to 100° C. Figure 4.6 shows the temperature profiles of the gearbox surface at a gear oil temperature of 100°C. Figure 4.6(a) shows the temperature profile of gear train assembly at $h_{100\ \text{W/m}^2\text{k}}$, the convective heat transfer coefficient (h) value. The minimum temperature is 342.81 K, and the maximum is 364.02 K. The maximum temperature effect is found on the same place as for $h_{100\ \text{W/m}^2\ \text{k}}$ at 80°C on the counter shaft right end in red hues. Red hues show hot areas subjected to thermal stresses and deformation. Figure 4.8, h_{100} shows the gradual increment in temperature profile of the gear train at different points. Figure 4.6(b) explains the temperature variations in the transmission gear train at $h_{200\ \text{W/m}^2\text{k}}$. As the value of h increases, there is an increase in temperature of the gear train by 4.76 K, noted by the red hues. The high hues temperature region is found at the fixed portion at the right side end of the countershaft. Figure 4.6(c) shows dark yellow hues on the third gear, designating an increment in temperature. The maximum value of temperature reached 370.09 K. The right-side fixed countershaft end temperature is the maximum for $h_{300\ \text{W/m}^2\text{k}}$. The temperature profile varies between 342.8 and 370.09 K. In Figure 4.8, h_{300} (3) shows the linear temperature variation at different points on the gear train. Figure 4.6(d) shows

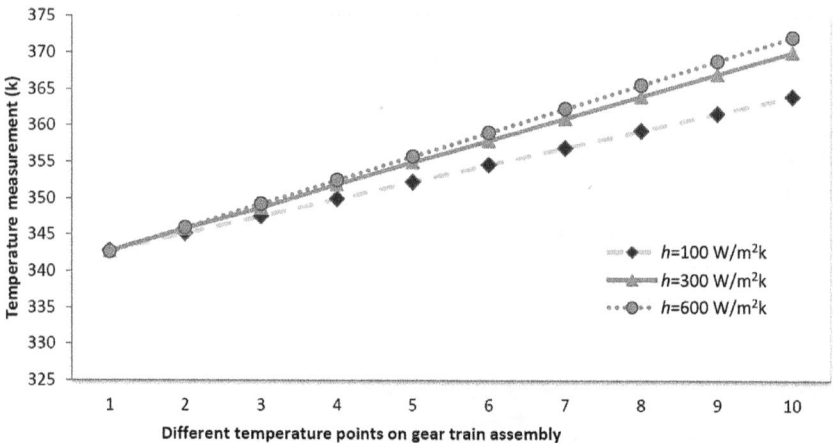

Figure 4.8 Temperature variations on gearbox surface at oil bath temperature of 100°C with varying h.

the gradual change in temperature. The profiles of the second and third gears show red hues of hot areas. In Figure 4.8, h_{600} shows the gear train temperature profile variation.

4.5 FEA RESULTS AND VALIDATION

Multispeed transmission gearbox and gear oil analysis is a highly nonlinear problem. From FEA analysis (Figures 4.5 and 4.7), when h is 100 W/m²k, and gear oil temperature is constant at 80°C, the gear train profile temperature varies very gradually and linearly. As we increase the value of convective heat transfer coefficient to 200 W/m² k, the difference in maximum temperature is only 1.58 K. Further, as there is an increment in the value of h (300, 400, 500, and 600), the difference in the gearbox surface temperature profile varies only 1.58–2.7 K (Figure 4.7).

The same temperature profile as in Figure 4.7 is generated when the gear oil temperature is constant at 100°C and the value of h increases from 100 to 600 W/m² k; the difference in temperature is 1–8 K (Figure 4.8). From Figures 4.7 and 4.8, it can be concluded that, if the gear oil temperature is constant (isothermal), the gear train thermal stresses at each convective heat transfer coefficient value are within permissible limits. The temperature profile of the gear train shows that the lower temperature is approximately constant (Figures 4.5 and 4.6) and that the maximum temperature varies by 1–5 K (Figures 4.7 and 4.8). The convective heat transfer coefficient h refers to the transfer of heat with fluid movement. In general, for the increase value of h, the rate of heat transfer increases. It signifies that the gearbox surface temperature should be reduced with an increase in the h value. The numerical simulation results show the same results as per the thermal concept. In this research work for thermal analysis, the gear oil temperature varied from 80°C to 100°C, and h varied 100–600 W/m²k; for these loading conditions, the temperatures at 10 different gear assembly points were measured.

The FEA-based numerical simulation result was validated with experimental results available in the literature (Figure 4.9). Long et al. [43] performed the experimental thermal analysis of a single gear tooth of a vehicle gearbox. The experimental study was performed at 2000, 6000, and 10000 rpm with combined loading. At different rotational speeds and loading conditions, the gear surface temperature was evaluated. The gear temperature is mainly governed by loading conditions, and rotational speed has less effect. In this study, when the vehicle was running at low speed (1500–2000 rpm) at 245 Nm load value, the gearbox surface temperature varied from 349.56 to 353.82 K, and the experimental results varied from 348.14 to 360.42 K. The FEA simulation temperature range lies within the experimental gear surface temperature. The deviation in results for minimum temperature is less than 1%, and for the maximum temperature, the distribution deviation is less than 2%. It indicates that deviation in result is

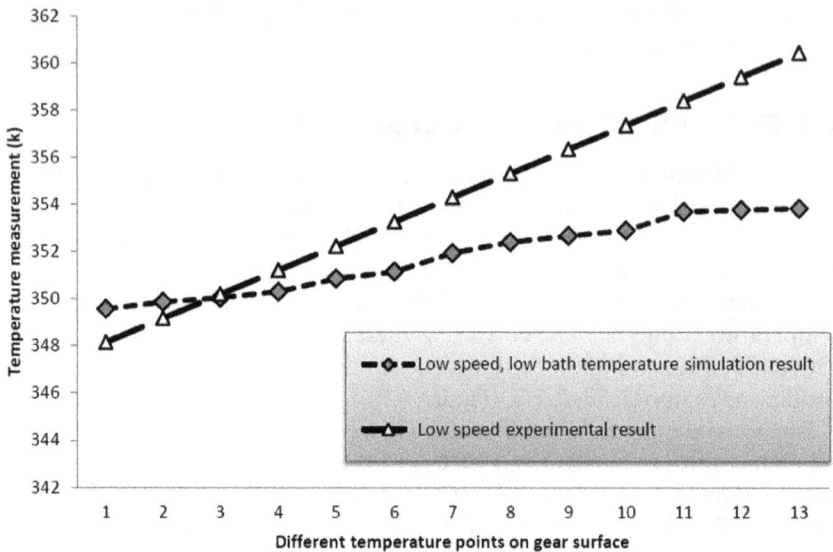

Figure 4.9 FEA simulation result validation using experimental results.

Source: Long et al. [43].

less than 5%, it shows the satisfactory measurement of the gearbox surface temperature.

4.6 CONCLUSIONS AND FUTURE SCOPE

The analysis of a multispeed transmission gearbox with gear oil has analytical and theoretical significance for the thermomechanical performance improvement of the gearbox. The FEA simulation results of a medium duty transmission gearbox concludes that:

1. A gradual increase in temperature around engaged gear pairs were reported due to the effect of heat generation on the surface of gears under the influence of gear meshing, frictional heat, average heat transfer, and gear oil bath temperature. The increment in temperature is very low because the resultant thermal stress generation due to the meshing action of the gear is much less.
2. It was found that varying the convective heat transfer (h) method reduces thermal stresses generation (Figures 4.5 and 4.6). All stresses, like equivalent von Mises stresses and thermal strain, are in the permissible range for the transmission gearbox.

3. The most important outcome of this analysis is that, by the varying convective heat transfer coefficient h, the phenomenon of overheated gear oil is less, which means that cooling will be greater and the thermomechanical performance of the multispeed transmission system will be enhanced.
4. An increase in the thermomechanical performance of a multispeed transmission system signifies higher fuel economy.

REFERENCES

[1] Dogan, S.N. (1999). Loose part vibration in vehicle transmissions Gear rattle. Transactions Journal of Engineering and Environmental Science, 23, 439–454.
[2] Wang, J., Zheng, J., & Yang, A. (2012). An Analytical Study of Bifurcation and Chaos in a Spur Gear Pair with Sliding Friction. Procedia Engineering, 31, 563–570.
[3] Abouel-Seoud, S.S., Mohamed, E.S., Abdel-Hamid, A.A., & Abdallah, A.S. (2013). Analytical Technique for Predicting Passenger Car Gearbox Structure Noise Using Vibration Response Analysis. British Journal of Applied Science & Technology, 3(4), 860–883.
[4] Vandi, G., Cavina, N., Corti, E., Mancini, G., Moro, D., Ponti, F., & Ravaglioli, V. (2014). Development of a software in the loop environment for automotive powertrain systems. Energy Procedia, 45, 789–798.
[5] Nacib, L., Pekpe, K.M., & Sakhara, S. (2013). Detecting gear tooth cracks using cepstral analysis in gearbox of helicopters. International Journal of Advances in Engineering & Technology, 5, 139–145.
[6] Gordon, T.J., & Bareket, Z. (2007). Vibration Transmission from Road Surface Features– Vehicle Measurement and Detection. Technical Report for Nissan Technical Center North America, Inc. UMTRI- 2007-4.
[7] Kar, C., & Mohanty, A.R. (2006). Monitoring gear vibrations through motor current signature analysis and wavelet transform. Mechanical Systems and Signal Processing, 20, 158–187.
[8] Czech, P. (2012). Diagnosis of industrial gearboxes Condition by vibration and time frequency, Scale-frequency, frequency-frequency analysis. Metalurgija, 51, 521–524.
[9] Singh, R. (2000). Dynamic design of automotive systems-Engine mounts and structural joints. Dynamic Design of Automotive Systems, 25, 319–330.
[10] Tuma, J. (2009). Gearbox Noise and Vibration Prediction and Control. International Journal of Acoustics and Vibration, 14, 1–11.
[11] Kumar, A., Jaiswal, H., Patil, P.P. 2014. "FEA Based Study of Loose Transmission Gearbox Housing" International Journal of Manufacturing, Materials, and Mechanical Engineering, Vol. 4(4), pp. 26–37. DOI:10.4018/ijmmme.2014100102.
[12] Yu, F., Li, Y., Sun, D., Shen, W., & Xia, W. (2013). Analysis for the Dynamic Characteristic of the Automobile Transmission Gearbox. Research Journal of Applied Sciences, Engineering and Technology, 5, 1449–1453.

[13] Kumar, A., Jaiswal, H., Patil, P.P. 2014. "FEM Simulation Based Computation of Natural Frequencies and Mode Shapes of Loose Transmission Gearbox Casing," International Review on Modelling and Simulations, Vol. 7 (5), pp. 900–905. DOI: http://dx.doi.org/10.15866/iremos.v7i5.3932.

[14] Kumar, A., Patil, P. 2017. "Free Vibration and Connecting Bolt Constraint Based FEA Analysis of Heavy Vehicle Medium Duty Transmission Gearbox Housing Made from AISI 4130 Alloy," Mathematics Applied to Engineering, Academic Press Elsevier, pp. 133–146, DOI: http://dx.doi.org/10.1016/B978-0-12-810998-4.00007-7, ISBN: 978-0-12-810998-4.

[15] Kumar, A., Jain, R., Patil, P.P. 2016. "Dynamic Analysis of Heavy Vehicle Medium Duty Drive Shaft Using Conventional and Composite Material" IOP: Materials Science and Engineering Vol. 149, pp. 1–11, DOI: 10.1088/1757-899X/149/1/012156.

[16] Kostic, S.C., & Ognjanovic, M. (2007). The Noise Structure of Gear Transmission Units and the Role of Gearbox Walls. FME Transactions, 35,105–112.

[17] Kumar, A., Jaiswal, H., Anand, A., Patil, P.P. 2014. "Mode Shape Vibration Analysis of Truck Transmission Housing Based on FEA," International Journal of Advancements in Mechanical and Aeronautical Engineering, Vol. 1(3), pp. 150–154. ISSN: 2372–4153.

[18] Shukla, A., Kumar, A. 2021, "Design and Thermal Characteristic Analysis of Gearbox System Based on Finite Element Analysis (FEA)" 3rd International Conference on "Advancements in Aeromechanical Materials for Manufacturing" AIP Conf. Proc. 2317, 030002-1–030002-5; https://doi.org/10.1063/5.0036235 Published by AIP Publishing. 978-0-7354-4058-6/$30.00.

[19] Patil, P.P., Sharma, S.C., Jaiswal, H., Kumar, A. 2014. "Modeling Influence of Tube Material on Vibration based EMMFS using ANFIS", Procedia Materials Science, Vol. 6, pp. 1097–1103. DOI: 10.1016/j.mspro.2014.07.181.

[20] Patil, P.P., Sharma, S.C., Saini, A., Kumar, A. 2014. "ANN Modelling of Cu Type Omega Vibration Based Mass Flow Sensor", Procedia Technology, Vol. 14, pp. 260–265. DOI: 10.1016/j.protcy.2014.08.034.

[21] Velex, P., & Flamand, L. (1996). Dynamic response of planetary trains to mesh parametric excitations. Journal of Mechanical Design Transactions of the ASME, 118, 7–14.

[22] Li, Y., & Guagnqiang, W. (2006). Analysis on Fatigue Life of Rear Suspension Based on Virtual Test rig. Computer Aided Engineering, 15, 128–130.

[23] Ansys R 14.5. (2013). Academic, Structural analysis Guide.

[24] Solidedge (2006). Version 19.0.

[25] Pro-E 5.0. (2013). Designing guide manual.

[26] Kumar, A., & Patil, P.P. (2016). Modal Analysis of Heavy Vehicle Truck Transmission Gearbox Housing Made from Different Materials. Journal of Engineering Science and Technology, Vol 11 (2), pp. 252–266.

[27] Patil, P.P., Kumar, A. 2016. "Dynamic Structural and Thermal Characteristics Analysis of Oil Lubricated Multi Speed Transmission Gearbox: Variation of Load, Rotational Speed and Convection Heat Transfer," Iranian Journal of Science and Technology Transactions of Mechanical Engineering (ISTM), pp. 1–11, DOI: 10.1007/s40997-016-0063-z.

[28] Kumar, A., Patil, P.P. 2016. "FEA Simulation and RSM Based Parametric Optimization of Vibrating Transmission Gearbox Housing" Perspectives in Science, Vol. 8 pp. 388–391 http://dx.doi.org/10.1016/j.pisc.2016.04.085.

[29] Kumar, A., Patil, P.P. 2016. "FEA Simulation Based Performance Study of Multi-Speed Transmission Gearbox," International Journal of Manufacturing, Materials, and Mechanical Engineering Vol. 6(1), pp. 57–67, DOI: 10.4018/IJMMME.2016010103

[30] Abbes, M.S., Fakhfakh, T., Haddar, M., & Maalej, A. (2007). Effect of transmission error on the dynamic behaviour of gearbox housing. International Journal of Advanced Manufacturing Technology, 34, 211–218.

[31] Marburg, S., & Hardtke, H.-J. (2001). Shape Optimization of a Vehicle Hat-Shelf: Improving Acoustic Properties for Different Load Cases by Maximizing First Eigen frequency, Computers & Structures, 79 (20–21), 1943–1957.

[32] Li, Z., & Liang, X. (2007). Vibro-Acoustic Analysis and Optimization of Damping Structure with Response Surface Method. Materials & Design, 28 (7), 1999–2007.

[33] Yuksel, E., Kamci, G., & Basdogan. I. (2012). Vibro-Acoustic Design Optimization Study to Improve the Sound Pressure Level Inside the Passenger Cabin. Journal of Vibration and Acoustics, 134 (6), 061017-1-061017-9. doi:10.1115/1.4007678.

[34] Marburg, S., Beer, H.-J., Gier, J., & Hardtke, H.-J. (2002). Experimental Verification of Structural Acoustic Modelling and Design Optimization. Journal of Sound and Vibration, 252(4), 591–615.

[35] Liang, X., Lin, Z., & Zhu, P. Acoustic Analysis of Damping Structure with Response Surface Method, Applied Acoustics, 68 (9), 1036–1053. doi:10.1016/j.apacoust.2006.05.021.

[36] Kumar, A., Joshi, H., Patil, P.P. (2014). "Vibration Based Failure Analysis of Heavy Vehicle Truck Transmission Gearbox Casing Using FEA," International Conference on Mechanical Engineering, pp. 251–259. ISBN: 9789-3510-72713.

[37] Ahmad, F., Tomer, V., Kumar, A., Patil, P.P. (2016). "FEA Simulation Based Thermo-Mechanical Analysis of Tractor Exhaust Manifold" Springer Book Series: Lecture Notes in Mechanical Engineering: CAD/CAM, Robotics and Factories of the Future, pp. 173–181 DOI: 10.1007/978-81-322-2740-3_18 ISBN: 978-81-322-2740-3.

[38] Hambric, S.A. (1995). Approximation Techniques for Broad-Band Acoustic Radiated Noise Design Optimization Problems. ASME Journal of Vibration Acoustics, 117(1), 136–144.doi:10.1115/1.2873857.

[39] Danti, M., Vige, D. & Nierop, G.V. (2010). Modal Methodology for the Simulation and Optimization of the Free-Layer Damping Treatment of a Car Body. ASME Journal of Vibration Acoustics, 132(2), 021001. doi: 10.1115/1.4000844

[40] Danti, M., Meneguzzo, M., Saponaro, R. & Kowarska, I. (2010). Multi-objective Optimization in Vehicle Concept Modeling," Proceedings of ISMA2010,USD2010, pp. 4095–4108.

[41] Kumar, Y.R., Rao, C.S.P., & Reddy, T.A. J. (2008). Robust process optimisation for fused deposition Modelling, International Journal of Manufacturing Technology and Management, 14, 228–245.

[42] Kumar, A., Jain, R., Jaiswal, H., Patil, P. 2015. "Dynamic Structure and Vibration Characteristics Analysis of Single Piece Drive shaft Using FEM," International Journal of Applied Engineering Research, Vol. 10 (11), pp. 10263–10269. ISSN: 0973-4562.

[43] Long, H., Lord, A. A., Gethin, D. T., & Roylance, B. J. (2003). Operating temperatures of oil-lubricated medium-speed gears: numerical models and experimental results, Proceedings of the I MECH E part G: Journal of aerospace engineering, 217 (2), 87–106. http://dx.doi.org/10.1243/0954410 03765208745.

Chapter 5

FEM-based design and free vibration analysis of composite plate

Harsh Kumar Bhardwaj and Vipin Kumar Sharma

5.1 INTRODUCTION

Composite materials are composed of mainly two materials: matrix and powder reinforcement (Sharma et al., 2019, Sharma et al., 2019). The purpose of the matrix is to offer a satisfactorily level of performance to the composite structure by holding the fibers jointly and transmitting the load to the fibers, which ultimately stops a crack from forming. Reinforcement helps in the load-carrying property of the composites (Sharma et al., 2019, Sharma et al., 2021). From stiffness to strength and thermal stability, almost every mechanical property results in the reinforcement type used in the composite materials. Another important category of composites is sandwich structure composite, which is widely used because its excellent properties make it favorable for every structural application. In this composite type, a thin layer of composite is attached to a lightweight and low-strength core. The core thickness provides a higher bending stiffness to the composite sandwich plate. Brischetto (2014) proposed a 3D free vibration analysis of multilayered structures based on a layer-wise approach. The free vibration analysis embedding isotropic and orthotropic materials are presented for several vibration modes, thickness ratios, imposed wave numbers, geometries, and multilayer configurations. Liu et al. (2016) analyzed the free vibration response of functionally graded and laminated composite sandwich shells using the differential quadrature finite element method (DQFEM). Hassaine et al. (2015) used B-spline finite strip element models to investigate the same. Phan-Dao et al. (2016) presented a generalized layer-wise higher-order shear deformation theory using improved mesh-free

DOI: 10.1201/9781003360001-5

radial point interpolation for static, free vibration, and buckling analysis of symmetrically laminated composite and sandwich plates.

Kant and Swaminathan (2001) presented a theoretical model for deformations of laminates that accounts for the effects of transverse shear deformation, transverse normal stress/strain under a nonlinear variation of in-plane displacements concerning thickness coordinates. Lurlalo et al. (2013) used zigzag theory with different boundary conditions to solve the problems of bending, free vibration, and buckling for multilayered plates. Subsequently, these equations are also specialized to problems. Natarajan and Manickam (2012) used first-order shear deformation theory (FSDT) to investigate sandwich functionally graded plates' bending and free vibration behavior. Gherlone et al. (2011) deal with two- and three-node beam finite elements to analyze laminated composite and sandwich beams. The refined zigzag theory (RZT) does not require shear correction factors as used in FSDT. Versino et al. (2013) developed six- and three-node C° continuous RZT-based triangular plate finite elements. Loja et al. (2015) used different shear deformation theories to formulate other layer-wise models, implemented through Kriging-based finite elements.

Khare et al. (2004) presented the free vibration analysis of isotropic, orthotropic, and layer-wise anisotropic composite and sandwich laminates using C° isoparametric formulation. Hwu et al. (2017) investigated the natural frequencies of sandwich plates and cylindrical shells composed of two composite laminated faces and ideally orthotropic elastic core using modified FSDT. Dehkordi et al. (2013) presented a new set of models in Carrera's unified formulation (CUF) framework for the static analysis of sandwich plates. For the accurate prediction of natural frequencies of thin and thick sandwich plates with functionally graded material (FGM) core, Dozio (2013) dealt with the formulation of advanced two-dimensional Ritz-based models. Postbuckling and postbuckled vibration analysis of sandwich plates subjected to nonuniform in-plane loading was investigated by Dey et al. (2016). The authors considered the core compressibility effects in the model by assuming fourth- and fifth-order expansions for the transverse and tangential displacement of the core. Birman and Kardomateas (2018) reviewed the trends in research and applications of sandwich structures, nanotubes, and intelligent materials. Li et al. (2016) extended the layer-wise/solid-element (LW/SE) method to the sandwich structure's static and free vibration analysis with the multilayered core. The authors employed eight-noded solid elements to discretize the cores and also used the layer-wise theory in the LW/SE method to model the behavior of laminated composite face sheets.

Neves et al. (2013) obtained governing equations and boundary conditions using the principle of virtual displacements under Carrera's unified formulation for free vibration response. Using a nine-node heterosis plate bending element based on first-order shear deformation theory with an a priori shear correction factor, Nayak and Satapathy (2016) carried out stochastic free vibration analysis of composite sandwich plates. Carbon-nanotube-reinforced composite (CNTRC) was investigated by Wang et al. (2018). The authors proposed a new polynomial refined plate theory for the core, and the classical plate theory was adopted for face sheets. Chakrabarti and Sheikh (2004) developed a six-noded triangular element based on refined plate theory with stiff laminated face sheets for free vibration analysis of sandwich plates. Using an improved discrete Kirchhoff quadrilateral element based on third-order shear deformation theory, Kulkarni and Kapuria (2008) studied the free vibration analysis of composite and sandwich plates. The effect of thermomechanical loads on sandwich plates has been analyzed by Shariyat (2010). The result of the thermal environment on laminated anisotropic face sheets has been visualized by Fazzolari and Carrera (2013).

Using Hill's criterion for orthotropic material in combination with Ilyushin's criterion for isotropic plates and shells, Chang et al. (2006) developed an approach for the elastoplastic analysis of sandwich panels with corrugated core. With the help of layer-wise theory, Momeni et al. (2017) developed a finite element model for a five-layer laminated composite beam. Two layers were filled with magneto-rheological fluid, and three layers were of composite materials. Jun et al. (2014) used Hamilton's principle to derive governing equations of motion for laminated shallow curved beams based on trigonometric shear deformation theory. After that, Jun et al. (2014) analyzed the free vibrations of the shallow curved beam using the dynamic stiffness method. Chang and Liu (2001) studied the vibration behavior of rectangular composite plates in contact with the fluid. Also, they analyzed the vibration behavior of simply supported and fully clamped composite plates with different widths and lengths. Jane and Harn (2000) studied the buckling and vibration analysis of a simply supported and fully attached composite orthotropic beam and plate with multidelamination using the reduced flexural stiffness method.

Based on the literature review, the study of sandwich plates using commercial software is limited in number. Hence, the present study aims to develop a FEM model for the sandwich plate. The study is extended to analyze the vibration behavior of sandwich plates with size ratio, thickness ratio, number of layers, angle of face sheet laminate, and boundary conditions.

5.2 MATERIALS AND METHODS

The finite element tools are implemented with the aim of free vibration analysis of sandwich plates. In addition, eight-node structural elements with 6 degrees of freedom at each node are suitable for analyzing thin to moderately thick face sheets. A higher-order 3D 20-node structural element with 3 degrees of freedom at each node is ideal for exploring the sandwich core.

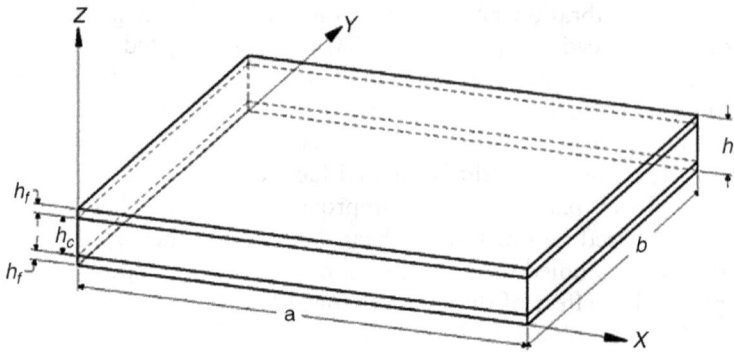

Figure 5.1 Geometrical representation of a sandwich plate.

Several shear deformation theories have been developed, classified mainly into three categories: classical plate theory (CPT), first-shear deformation theory (FSDT), and higher-order shear deformation theory. The first such shear deformation theory was proposed by Stavski (1965). Recently Thai and Choi (2013) developed a first-order shear deformation theory.

Our study considers a sandwich plate of thickness h comprised of face sheet thickness h_f and core thickness h_c, as shown in the Figure 5.1.

The transverse displacement w is comprised of two components of bending (w_b) and shear (w_s), respectively.

$$w(x,y,z) = w_b(x,y) + w_s(x,y)$$

The displacement component in the x-direction (u) and the y-direction (v) comprises extension, bending, and shear components respectively:

$$u = u_0 + u_b + u_s, \quad v = v_0 + v_b + v_s$$

According to classical plate theory (CPT), the expression for bending components u_b and v_b is given by

$$u_b = -z\frac{\partial w_b}{\partial x}, \quad v_b = -z\frac{\partial w_b}{\partial y}$$

And the expression for shear components u_s and v_s can be given as

$$u_s = -f(z)\frac{\partial w_s}{\partial x}, \quad v_s = -f(z)\frac{\partial w_s}{\partial y}$$

Shape function $f(z)$ has been defined by Karama et al. (2003) as

$$f(z) = z - ze^{-2\frac{z^2}{h^2}}$$

The displacement fields of the existing FSDT are given by

$$u(x,y,z) = u_0(x,y) - z\frac{\partial w_b}{\partial x} - f(z)\frac{\partial w_s}{\partial x}$$

$$v(x,y,z) = v_0(x,y) - z\frac{\partial w_b}{\partial y} - f(z)\frac{\partial w_s}{\partial y} \qquad (5.1)$$

$$w(x,y,z) = w_b(x,y) + w_s(x,y)$$

where u_0 and v_0 are the displacements of the plate about the midplane in $x-$ and y-directions, respectively. w_b and w_s are the bending and shear components of transverse displacement, respectively, and $f(z)$ is the shape function to determine the transverse shear stress and shear strain distribution along with the thickness of the plate, which ensures zero transverse shear stress at the top and bottom of the plate. The shape function has two expressions according to hyperbolic shear deformation theory (HSDT) proposed by Hassaine et al. (2012, 2015) and according to exponential shear deformation theory (ESDT) proposed by Karama et al. (2003).

The nonzero strains associated with the displacement field (equation [5.1]) are as follows

$$\epsilon_x = \frac{\partial u_0}{\partial x} - z\frac{\partial^2 w_b}{\partial x^2} - f(z)\frac{\partial^2 w_s}{\partial x^2} \qquad (5.2a)$$

$$\epsilon_y = \frac{\partial v_0}{\partial y} - z\frac{\partial^2 w_b}{\partial y^2} - f(z)\frac{\partial^2 w_s}{\partial y^2} \qquad (5.2b)$$

$$\gamma_{xy} = \frac{\partial u_0}{\partial y} + \frac{\partial v_0}{\partial x} - 2z\frac{\partial^2 w_b}{\partial x \partial y} - 2f(z)\frac{\partial^2 w_s}{\partial x \partial y} \qquad (5.2c)$$

$$\gamma_{xz} = \frac{\partial w_s}{\partial x} \qquad (5.2d)$$

$$\gamma_{yz} = \frac{\partial w_s}{\partial y} \qquad (5.2e)$$

According to Hook's law, for each layer of the plate, the state of stress is given by

$$
\begin{bmatrix} \sigma_x \\ \sigma_y \\ \tau_{xy} \\ \tau_{yz} \\ \tau_{xz} \end{bmatrix} = \begin{bmatrix} P_{11} & P_{12} & 0 & 0 & 0 \\ P_{12} & P_{22} & 0 & 0 & 0 \\ 0 & 0 & P_{66} & 0 & 0 \\ 0 & 0 & 0 & P_{44} & 0 \\ 0 & 0 & 0 & 0 & P_{55} \end{bmatrix} \begin{bmatrix} \epsilon_x \\ \epsilon_y \\ \gamma_{xy} \\ \gamma_{yz} \\ \gamma_{xz} \end{bmatrix} \tag{5.3a}
$$

where P_{ij} refers to the material constants in the material axes of the layer, given as

$$
P_{11} = \frac{E_1}{1 - \vartheta_{12}\vartheta_{21}}, \quad P_{12} = \frac{\vartheta_{12}E_2}{1 - \vartheta_{12}\vartheta_{21}}, \quad P_{22} = \frac{E_2}{1 - \vartheta_{12}\vartheta_{21}}, \quad P_{66} = G_{12},
$$

$$
P_{44} = G_{23}, \quad P_{55} = G_{13} \tag{5.3b}
$$

The stress–strain relations in the laminate coordinates of a kth layer are

$$
\begin{bmatrix} \sigma_x \\ \sigma_y \\ \tau_{xy} \\ \tau_{yz} \\ \tau_{xz} \end{bmatrix}^{(k)} = \begin{bmatrix} \overline{P_{11}} & \overline{P_{12}} & \overline{P_{16}} & 0 & 0 \\ \overline{P_{12}} & \overline{P_{22}} & \overline{P_{26}} & 0 & 0 \\ \overline{P_{16}} & \overline{P_{26}} & \overline{P_{66}} & 0 & 0 \\ 0 & 0 & 0 & \overline{P_{44}} & \overline{P_{45}} \\ 0 & 0 & 0 & \overline{P_{45}} & \overline{P_{55}} \end{bmatrix}^{(k)} \begin{bmatrix} \epsilon_x \\ \epsilon_y \\ \gamma_{xy} \\ \gamma_{yz} \\ \gamma_{xz} \end{bmatrix} \tag{5.3c}
$$

where $\overline{P_{ij}}$, the material transformation constants, are given by Karama et al. (2003) as

$$
\overline{P_{11}} = P_{11}\cos^4\theta + 2(P_{12} + 2P_{66})\sin^2\theta + P_{22}\sin^4\theta
$$

$$
\overline{P_{12}} = (P_{11} + P_{22} - 4P_{66})\sin^2\theta\cos^2\theta + P_{12}(\sin^4\theta + \cos^4\theta)
$$

$$
\overline{P_{22}} = P_{11}\sin^4\theta + 2(P_{12} + 2P_{66})\sin^2\theta\cos^2\theta + P_{22}\cos^4\theta
$$

$$\overline{P_{16}} = (P_{11} - P_{12} - 2P_{66})\sin\theta\cos^3\theta + (P_{12} - P_{22} + 2P_{66})\sin^3\theta\cos\theta$$

$$\overline{P_{26}} = (P_{11} - P_{12} - 2P_{66})\sin^3\theta\cos\theta + (P_{12} - P_{22} + 2P_{66})\sin\theta\cos^3\theta$$

$$\overline{P_{66}} = (P_{11} + P_{22} - 2P_{12} - 2P_{66})\sin^2\theta\cos^2\theta + P_{66}(\sin^4\theta + \cos^4\theta)$$

$$\overline{P_{44}} = P_{44}\cos^2\theta + P_{55}\sin^2\theta$$

$$\overline{P_{45}} = (P_{55} - P_{44})\cos\theta\sin\theta$$

$$\overline{P_{55}} = P_{55}\cos^2\theta + P_{44}\sin^2\theta \tag{5.3d}$$

where θ is the angle between the global x-axis and the local x-axis.

To derive the equation of motion of laminated plate, Hamilton principle is used as follows:

$$\int_0^T (\delta U - \delta W - \delta K)\,dt = 0 \tag{5.4}$$

where $\delta U, \delta W$, and δK are the variations of strain energy, work done, and kinetic energy, respectively, of the plate. Taking the variation of Hamilton's analytical equation and integrating by parts, the variation of strain energy is obtained as

$$\delta U = \int_A \int_{-h/2}^{h/2} \left(\sigma_x \delta\varepsilon_x + \sigma_y \delta\varepsilon_y + \sigma_{xy}\delta\gamma_{xy} + \sigma_{xz}\delta\gamma_{xz} + \sigma_{yz}\delta\gamma_{yz}\right) dA\,dz =$$

$$\int_A [N_x \frac{\partial \delta u_0}{\partial x} - M_x \frac{\partial^2 \delta w_b}{\partial x^2} + N_y \frac{\partial \delta v_0}{\partial y} - M_y \frac{\partial^2 \delta w_b}{\partial y^2} +$$

$$N_{xy}\left(\frac{\partial \delta u_0}{\partial y} + \frac{\partial \delta v_0}{\partial x}\right) - 2M_{xy}\frac{\partial^2 \delta w_b}{\partial x\partial y} + P_x \frac{\partial \delta w_s}{\partial x} + P_y \frac{\partial \delta w_s}{\partial y}]dA \tag{5.5}$$

$$(N_x, N_y, N_{xy}) = \int_{-h/2}^{h/2} (\sigma_x, \sigma_y, \sigma_{xy})\,dz \tag{5.6a}$$

$$(M_x, M_y, M_{xy}) = \int_{-h/2}^{h/2} (\sigma_x, \sigma_y, \sigma_{xy})z\,dz \tag{5.6b}$$

$$(P_x, P_y) = \int_{-h/2}^{h/2} (\sigma_{xz}, \sigma_{yz})\,dz \tag{5.6c}$$

Inserting the equations (5.2a), (5.2b), (5.2c), (5.2d), and (5.2e) into equation (5.3c) and the subsequent results into equations (5.6a), (5.6b), and (5.6c) gives

$$
\begin{bmatrix} N_x \\ N_y \\ N_{xy} \end{bmatrix} = \begin{bmatrix} B_{11} & B_{12} & B_{16} \\ B_{12} & B_{22} & B_{26} \\ B_{16} & B_{26} & B_{66} \end{bmatrix} \begin{bmatrix} \dfrac{\partial u_0}{\partial x} \\ \dfrac{\partial v_0}{\partial y} \\ \dfrac{\partial u_0}{\partial y} + \dfrac{\partial v_0}{\partial x} \end{bmatrix} + \begin{bmatrix} C_{11} & C_{12} & C_{16} \\ C_{12} & C_{22} & C_{26} \\ C_{16} & C_{26} & C_{66} \end{bmatrix} \begin{bmatrix} -\dfrac{\partial^2 w_b}{\partial x^2} \\ -\dfrac{\partial^2 w_b}{\partial y^2} \\ -2\dfrac{\partial^2 w_b}{\partial x \partial y} \end{bmatrix}
\tag{5.7a}
$$

$$
\begin{bmatrix} M_x \\ M_y \\ M_{xy} \end{bmatrix} = \begin{bmatrix} C_{11} & C_{12} & C_{16} \\ C_{12} & C_{22} & C_{26} \\ C_{16} & C_{26} & C_{66} \end{bmatrix} \begin{bmatrix} \dfrac{\partial u_0}{\partial x} \\ \dfrac{\partial v_0}{\partial y} \\ \dfrac{\partial u_0}{\partial y} + \dfrac{\partial v_0}{\partial x} \end{bmatrix} + \begin{bmatrix} D_{11} & D_{12} & D_{16} \\ D_{12} & D_{22} & D_{26} \\ D_{16} & D_{26} & D_{66} \end{bmatrix} \begin{bmatrix} -\dfrac{\partial^2 w_b}{\partial x^2} \\ -\dfrac{\partial^2 w_b}{\partial y^2} \\ -2\dfrac{\partial^2 w_b}{\partial x \partial y} \end{bmatrix}
\tag{5.7b}
$$

$$
\begin{bmatrix} P_y \\ P_x \end{bmatrix} = k \begin{bmatrix} B_{44} & B_{45} \\ B_{45} & B_{55} \end{bmatrix} \begin{bmatrix} \dfrac{\partial w_s}{\partial y} \\ \dfrac{\partial w_s}{\partial x} \end{bmatrix}
\tag{5.7c}
$$

where k is the shear correction factor, and $\left(B_{ij}, C_{ij}, D_{ij} \right)$ are the stiffness coefficients defined by

$$
\left(B_{ij}, C_{ij}, D_{ij} \right) = \int_{-h/2}^{h/2} \overline{P}_{ij} \left(1, z, z^2 \right) dz
\tag{5.8}
$$

The variation of work done by the transverse loads q can be written as

$$
\delta W = -\int_A q\,\delta\left(w_b + w_s \right) dA
\tag{5.9}
$$

The variation of kinetic energy can be expressed as

$$
\delta K = \int_V \left(\dot{u}_1 \delta \dot{u}_1 + \dot{u}_2 \delta \dot{u}_2 + \dot{u}_3 \delta \dot{u}_3 \right) \rho\, dA\, dz
$$

$$= \int_A \left\{ I_0 \left[\dot{u}_0 \delta \dot{u}_0 + \dot{v}_0 \delta \dot{v}_0 + \left(\dot{w}_b + \dot{w}_s \right) \delta \left(\dot{w}_b + \dot{w}_s \right) \right] + \right.$$

$$\left. I_2 \left(\frac{\partial \dot{w}_b}{\partial x} \frac{\partial \delta \dot{w}_b}{\partial x} + \frac{\partial \dot{w}_b}{\partial y} \frac{\partial \delta \dot{w}_b}{\partial y} \right) \right\} dA \tag{5.10}$$

where dot superscript means differentiation concerning the time, ρ is the mass density, and (I_0, I_2) are mass inertias defined by

$$\left(I_0, I_2 \right) = \int_{-b/2}^{b/2} \left(1, z^2 \right) \rho dz \tag{5.11}$$

Substituting the expression of δU, δW, δK from equations (5.5), (5.9), and (5.10), respectively, into equation (5.4) and integrating by parts, the following expressions for the equation of motions are obtained:

$$\delta u_0 : \frac{\partial N_x}{\partial x} + \frac{\partial N_{xy}}{\partial y} = I_0 \ddot{u}_0 \tag{5.12a}$$

$$\delta v_0 : \frac{\partial N_y}{\partial y} + \frac{\partial N_{xy}}{\partial x} = I_0 \ddot{v}_0 \tag{5.12b}$$

$$\delta w_b : \frac{\partial^2 M_x}{\partial x^2} + 2 \frac{\partial^2 M_{xy}}{\partial x \partial y} + \frac{\partial^2 M_y}{\partial y^2} + q = I_0 \left(\ddot{w}_b + \ddot{w}_s \right) - I_2 \nabla^2 \ddot{w}_b \tag{5.12c}$$

$$\delta w_s : \frac{\partial P_x}{\partial x} + \frac{\partial P_y}{\partial y} + q = I_0 \left(\ddot{w}_b + \ddot{w}_s \right) \tag{5.12d}$$

The natural boundary conditions are of the following form:

Clamped edge

$$u_0 = v_0 = w_b = w_s = \frac{\partial w_b}{\partial x} = 0, \text{ at } x = 0, a \tag{5.13a}$$

$$u_0 = v_0 = w_b = w_s = \frac{\partial w_b}{\partial y} = 0, \text{ at } y = 0, b \tag{5.13b}$$

Simply supported edge (cross-ply)

$$v_0 = N_x = w_b = w_s = M_x = 0, \text{ at } x = 0.a \tag{5.14a}$$

$$u_0 = N_y = w_b = w_s = M_y = 0, \text{ at } y = 0, b \tag{5.14b}$$

Simply supported edge (angle-ply)

$$v_0 = N_{xy} = w_b = w_s = M_x = 0, \text{ at } x = 0, a \tag{5.15a}$$

$$u_0 = N_{xy} = w_b = w_s = M_y = 0, \text{ at } y = 0, b \tag{5.15b}$$

5.3 NUMERICAL RESULTS AND DISCUSSIONS

The present investigation has been focused on the free vibration behavior of the sandwich plate. To analyze the sandwich plate with laminated cross-ply face sheets quantitatively, the following material properties have been used.

Face sheets:
$E_1 = 276$ GPa, $E_2 = E_3 = 6.9$ GPa
$G_{12} = G_{13} = G_{23} = 6.9$ GPa
$v_{12} = v_{13} = 0.25$, $v_{23} = 0.3$
$\rho = 681.8$ kg/m^3

Core:
$E_1 = E_2 = E_3 = 0.5776$ GPa
$G_{12} = G_{13} = 0.1079$ GPa, $G_{23} = 0.22215$ GPa
$v_{12} = v_{13} = v_{23} = 0.0025$
$\rho = 1000$ kg/m^3

Table 5.1 contains a list of acronyms used to denote the different plate theories.

Table 5.2 shows the convergence of nondimensional frequencies for a simply supported sandwich plate having two laminated layers on each face. As seen from Table 5.2, natural frequency variation converges for mesh size 21×21. To check the accuracy of the results obtained by current FEM models, the results are compared with those of Chakrabarti and Sheikh

Table 5.1 List of acronyms used to denote different plate theories

Acronym	Description
ZZF	Zig-zag FEM by Chakrabarti and Sheikh (2004)
3D	3D exact elasticity solution by Kulkarni and Kapuria (2008)
ZZA	Zigzag analytical by Kulkarni and Kapuria (2008)
RTO	Refined third order by Kulkarni and Kapuria (2008)
HSDT L-G	HSDT local–global by Shariyat (2010)

Table 5.2 Nondimensional frequencies $\bar{\omega} = 100\omega a\sqrt{\dfrac{\rho^c}{E_1^f}}$ convergence of a simply

supported antisymmetric (0°/90°/core/90°/0°) square sandwich plate with
$a/b = 1$, $a/h = 10$, and $h_f/h = 0.1$

Mode	Mesh size					
	11×11	13×13	15×15	17×17	19×19	21×21
1	9.386	9.386	9.386	9.386	9.386	9.386
2	16.075	16.074	16.074	16.073	16.073	16.073
3	21.400	21.399	21.398	21.398	21.398	21.398
4	25.791	25.785	25.782	25.781	25.780	25.779
5	25.792	25.791	25.790	25.790	25.790	25.790
6	33.043	33.043	33.043	33.043	33.043	33.043
7	33.151	33.147	33.145	33.145	33.144	33.144
8	33.540	33.540	33.540	33.540	33.540	33.540
9	36.668	36.640	36.627	36.620	36.616	36.614
10	39.030	39.015	39.007	39.004	39.004	39.004

Table 5.3 Comparative study of nondimensional frequencies $\bar{\omega} = 100\omega a\sqrt{\dfrac{\rho^c}{E_1^f}}$ of a simply

supported square sandwich plate with lamination scheme (0°°/90°/core/90°/0°)
with $a/b = 1$, $a/h = 20$, and $h_f/h = 0.1$

Mode	Present	3D FEM ANSYS (2013)	HSDT L-G (2010)	3D (2008)	ZZA (2008)	ZIGT FE (2008)	ZIGT FE (2004)
1	7.1915	7.825	7.6876	7.6882	7.689	7.681	7.929
2	11.1055	13.0503	13.8342	13.8455	13.8475	13.826	13.045
3	15.9941	17.3013	15.9188	15.9204	15.924	15.903	17.32
4	18.3514	19.0819	19.6407	19.6563	19.66	19.588	18.832
5	18.7735	20.2218	20.6683	20.676	20.6805	20.654	20.097
6	24.319	24.5643	24.9375	24.9485	24.954	24.842	24.143
Average error (%)		−7.09%	−7.32%	−7.36%	−7.38%	−7.17%	−6.71%

(2004), Kulkarni and Kapuria (2008), Shariyat (2010), and Fazzolari and Carrera (2013). With the refinement of mesh size, a good agreement of results is obtained between the present and previous work, as seen from Tables 5.3, 5.4, and 5.5.

Figure 5.2 shows the natural frequency variation for the first 10 modes of vibration for a simply supported sandwich plate with lamination scheme (0°/90°/core/90°/0°) with the aspect ratio or size ratio ($b/a = 1, 2, 3, 4$). It is clear from Figure 5.2 that the natural frequency decreases with an increase

Table 5.4 Comparative study of nondimensional frequencies $\bar{\omega} = 100\omega a\sqrt{\dfrac{\rho^c}{E_1^f}}$ of all sides

clamped square sandwich plate with (0°/90°/core/90°/0°) with $a/b = 1$, $a/h = 20$ and $h_f/h = 0.1$

Mode	Present	HSDT L-G (2010)	3D FE ABAQUS (2008)	ZIGT FE (2008)	ZIGT FE (2004)
1	9.7299	10.2815	10.1635	10.2954	10.536
2	14.347	15.0504	15.2734	15.5269	14.709
3	19.3234	17.1346	17.2645	17.5849	18.708
4	21.7374	20.5997	20.7882	21.2667	20.182
5	22.5505	21.6902	21.826	22.2081	21.369
6	28.2625	25.879	26.0346	26.695	25.406
Average error (%)		3.57%	3.01%	1.07%	2.94%

Table 5.5 Comparative study of nondimensional frequencies $\bar{\omega} = 100\omega a\sqrt{\dfrac{\rho^c}{E_1^f}}$ of a square

sandwich plate with lamination scheme (0°/90°/core/90°/0°) with $a/b = 1$, $a/h = 20$ and $h_f/h = 0.1$, SCSC

Mode	Present	HSDT L-G (2010)	3D FE ABAQUS (2008)	ZIGT FE (2008)	ZIGT FE (2004)
1	8.9456	8.6114	8.5675	8.6312	8.623
2	12.2234	14.1833	14.3524	14.5453	13.533
3	18.9743	16.2974	16.3014	16.3462	17.601
4	19.0159	19.2974	20.0117	20.1847	19.441
5	21.4001	20.9096	21.2553	21.5761	20.416
6	26.4275	24.9557	25.2762	25.3159	24.62
Average error (%)		1.83%	0.54%	0.26%	1.97%

in aspect ratio. Figure 5.3 shows the variation of natural frequency for the first 10 modes of vibration for a simply supported square sandwich plate ($b/a = 1$) with a lamination scheme (0°/90°/core/90°/0°) with thickness ratio ($b/a = 0.05, 0.1, 0.15, 0.2$) of the plate. It is clear from the figure that the natural frequency decreases with an increase in the thickness ratio. Figure 5.4 shows the variation of natural frequency for the first 10 modes of vibration for a simply supported square sandwich plate ($b/a=1$) with lamination scheme (0°/90°/core/90°/0°) with face-sheet-thickness-to-plate thickness ratio ($h_f/h = 0.05, 0.1, 0.15$). It is clear from the figure that the natural frequency increases with face sheet to plate thickness.

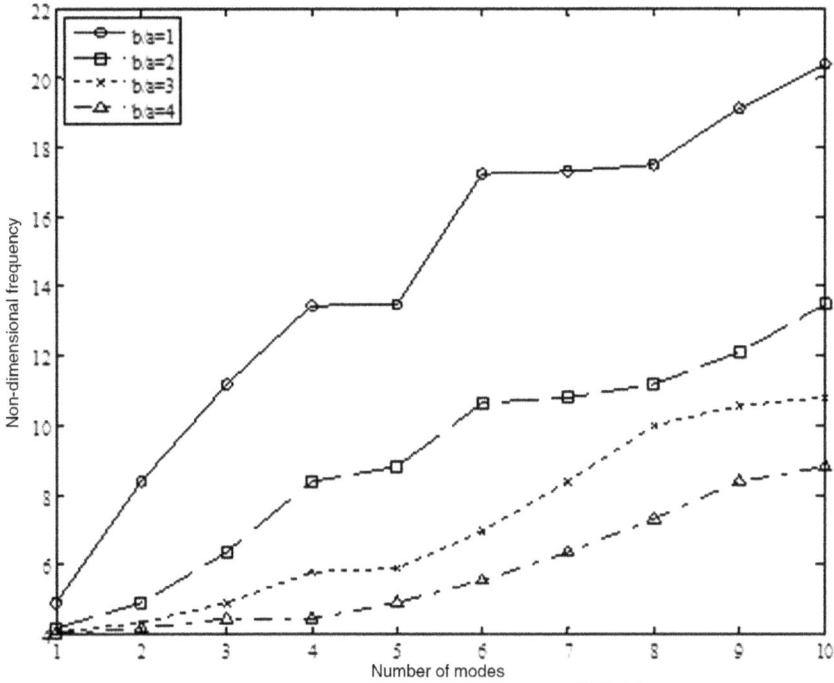

Figure 5.2 Variation of nondimensional frequencies $\bar{\omega} = \dfrac{\omega a^2}{h} \sqrt{\left(\dfrac{\rho}{E_2}\right)_f}$ for first 10 modes of vibration for a simply supported sandwich plate with lamination scheme (0°/90°/core/90°/ 0°) with size ratio (b/a = 1, 2, 3, 4) having thickness ratio h/a = 0.1 and face-sheet-to-plate-thickness ratio of h_f/h = 0.1.

Table 5.6 shows the natural frequency variation for the first 10 modes of vibration for a simply supported square sandwich plate (b/a = 1) with several face sheets (n = 2, 4, 6). As seen from the table, the natural frequency increases with an increase in the number of layers of the face sheet of the sandwich plate. Table 5.7 shows the natural frequency variation for the first 10 modes of vibration for a simply supported square sandwich plate (b/a = 1) with face sheet angle. It is clear from the table that natural frequency decreases with an increase in face sheet angle. Table 5.8 shows the variation of natural frequency for the first 10 modes of vibration for a square sandwich plate (b/a = 1) with lamination scheme (0°/90°/core/90°/0°) for different boundary conditions like all sides clamped (CCCC), all sides simply supported (SSS), any two consecutive sides simply supported, and the other two consecutive sides fully clamped (SSCC). Alternate sides are

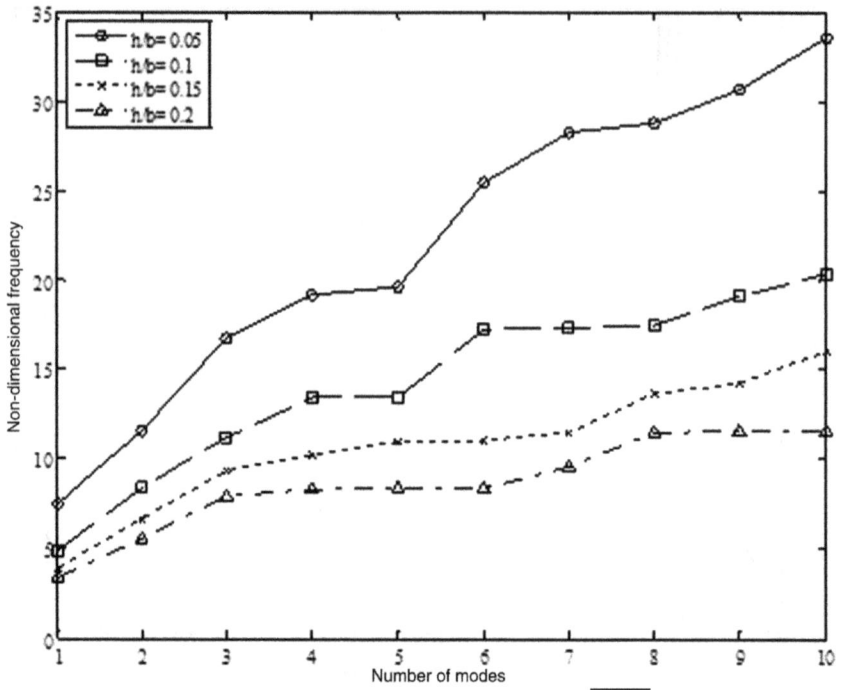

Figure 5.3 Variation of nondimensional frequencies $\bar{\omega} = \dfrac{\omega a^2}{h} \sqrt{\left(\dfrac{\rho}{E_2}\right)_f}$ for first 10 modes of vibration for a simply supported sandwich plate with lamination scheme (0°/90°/core/90°/0°) with thickness ratio (h/a = 0.05, 0.1, 0.15, 0.2), having size ratio b/a = 1 and face-sheet-to-plate-thickness ratio h_f/h = 0.1.

simply supported and fully clamped (SCSC). It is clear from the table that the natural frequency is maximum for all sides fully clamped (CCCC) and minimum for all sides simply supported (SSSS).

5.4 CONCLUSIONS AND FUTURE SCOPE

In this chapter, a FEM model is developed. The previously published results validate the accuracy of the results obtained by this model. It has been observed that with the refinement of mesh size, a good agreement of results is obtained between the present and the previously published work. Furthermore, the present study results conclude that the natural frequency of a simply supported sandwich plate increases with an increase in size ratio, face-sheet-to–plate-thickness ratio, and several layers. The natural frequency decreases with an increase in the thickness ratio of the plate and angle of

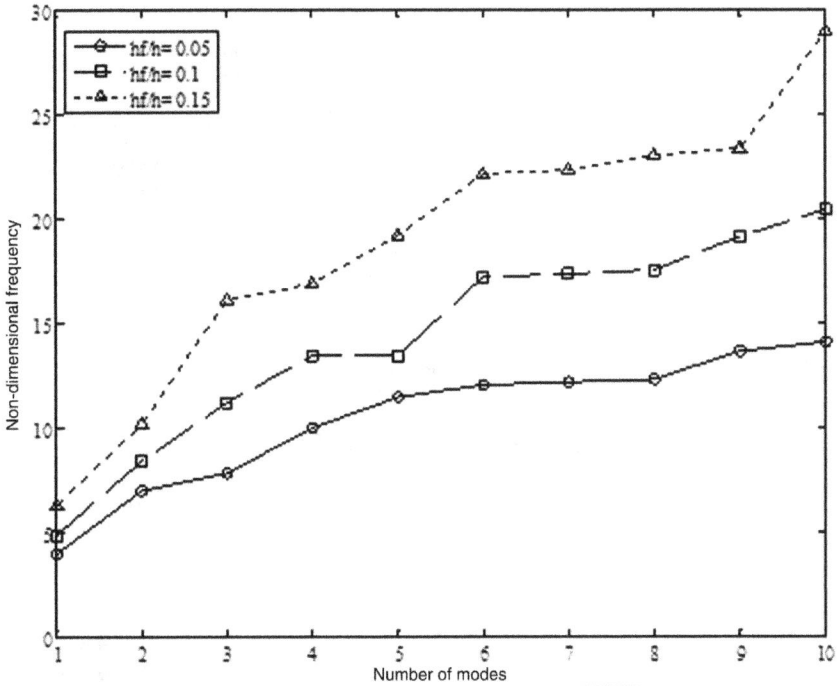

Figure 5.4 Variation of nondimensional frequencies $\bar{\omega} = \dfrac{\omega a^2}{h}\sqrt{\left(\dfrac{\rho}{E_2}\right)_f}$ for first 10 modes of vibration for a simply supported sandwich plate with lamination scheme (0°/90°/core/90°/0°) with face-sheet-to-plate-thickness ratio of $h_f/h = 0.05, 0.e1, 0.15$, having size ratio $b/a = 1$ and thickness ratio $h/a = 0.1$.

Table 5.6 Variation of nondimensional frequencies $\bar{\omega} = \dfrac{\omega a^2}{h}\sqrt{\left(\dfrac{\rho}{E_2}\right)_f}$ for first 10 modes of vibration for a simply supported sandwich plate with several layers (90°/core/90°, 0°/90°/core/90°/0°, 90°/0°/90°/core/90°/0°/90°) having size ratio $b/a = 1$, thickness ratio $h/a = 0.1$ and face-sheet-to-plate thickness ratio $h_f/h = 0.1$

Mode	No. of layers		
	90°/core/90°	0°/900/core/90°/0°	90°/0°/90°/core/90°/0°/90°
1	4.006	4.899	6.316
2	6.972	8.391	10.161
3	7.822	11.172	16.088
4	10.012	13.459	16.877
5	11.431	13.463	19.172

(continued)

Table 5.6 Cont.

Mode	90°/core/90°	0°/900/core/90°/0°	90°/0°/90°/core/90°/0°/90°
		No. of layers	
6	12.014	17.226	22.104
7	12.188	17.302	22.282
8	12.294	17.500	23.032
9	13.619	19.116	23.319
10	14.059	20.364	28.969

Table 5.7 Variation of nondimensional frequencies $\bar{\omega} = \dfrac{\omega a^2}{h}\sqrt{\left(\dfrac{\rho}{E_2}\right)_f}$ for first 10 modes

of vibration for a simply supported sandwich plate with the angle of face sheet lamina (0°/60°/core/60°/0°,0°/90°/core/90°/0°) having size ratio b/a = 1, thickness ratio , $/a$ = 0.1, and face-sheet-to plate-thickness ratio h_f/h = 0.1

Mode	0°/60°/core/60°/0°	0°/90°/core/90°/0°
	Angle of face sheet lamina	
1	4.902	4.899
2	8.394	8.391
3	11.174	11.172
4	13.462	13.459
5	13.468	13.463
6	17.255	17.226
7	17.308	17.302
8	17.515	17.500
9	19.118	19.116
10	20.366	20.364

Table 5.8 Variation of nondimensional frequencies $\bar{\omega} = \dfrac{\omega a^2}{h}\sqrt{\left(\dfrac{\rho}{E_2}\right)_f}$ for first 10 modes

of vibration for a sandwich plate with lamination scheme (0°/90°/core/90°/0°) with different boundary conditions (SSSS, CCCC, SSCC, SCSC) having size ratio b/a = 1, thickness ratio h/a = 0.1, and face-sheet-to-plate-thickness ratio h_f/h = 0.1

Mode	SSSS	CCCC	SSCC	SCSC
		Boundary condition		
1	4.899	7.067	5.887	6.468
2	8.391	10.451	9.383	9.397
3	11.172	14.441	12.739	14.123

Table 5.8 Cont.

Mode	SSSS	Boundary condition		
		CCCC	SSCC	SCSC
4	13.459	15.241	14.739	14.242
5	13.463	16.641	14.318	16.111
6	17.226	20.208	18.701	17.515
7	17.302	20.745	19.656	19.443
8	17.500	24.480	19.904	19.600
9	19.116	24.794	22.400	23.862
10	20.364	25.055	23.421	24.393

face sheet lamina. Natural frequency is maximum for an all-side clamped sandwich plate and minimum for all sides simply supported sandwich plate.

REFERENCES

Birman, V. & Kardomateas, G.A. (2018). Review of current trends in research and applications of sandwich structures. Composites: Part B, 142, 221–240.

Brischetto, S. (2014). An exact 3D solution for free vibrations of multilayered cross-ply composite and sandwich plates and shells. International Journal of Applied Mechanics, 6 (6).

Chakrabarti, A., & Sheikh, A.H. (2014). The vibration of laminate-faced sandwich plate by a new refined element. Journal of Aerospace Engineering, 17, 123–134.

Chang, T.P., & Liu, M.F. (2001).Vibration analysis of rectangular composite plates in contact with fluid. Mechanics of Structures and Machines, 29(1), 101–120.

Chang, W.S., Krauthammer, T., & Ventsel, E. (2006). Elasto-plastic analysis of corrugated-core sandwich plates. Mechanics of Advanced Materials& Structures, 13, 151–160.

Dehkordi, M.B., Cinefra, M., Khalili, S.M.R., & Carrera, E. (2013). Mixed LW/ESL models for the analysis of sandwich plates with composite faces. Composite Structures, 98, 330–339.

Dey, T., Kumar, R., & Panda, S.K. (2016). Postbuckling and postbuckled vibration analysis of sandwich plates under non-uniform mechanical edge loadings. International Journal of Mechanical Sciences, 115, 226–237.

Dozio, L. (2013). Natural frequencies of sandwich plates with FGM core via variable-kinematic 2-D Ritz models. Composite Structures, 96, 561–568.

Fazzolari, F. A., & Carrera, E. (2013). Free vibration analysis of sandwich plates with anisotropic face sheets in thermal environment by using the hierarchical trigonometric Ritz formulation. Composites: Part B, 50, 67–81.

Gherlone, M., Tessler, A., & Di-Sciuva, M. (2011). C^0 beam elements based on the refined Zigzag theory for multilayered composite and sandwich laminates. Composite Structures, 93, 2882–2894.

Hassaine, D.T., Tounsi, T., & Hadgi L. (2012).A theoretical analysis for static and dynamic behaviour of functionally graded plates. Materials Physics and Mechanics, 2(2), 110–128.

Hassaine, D.T., Tounsi, T., &Hadgi, L. (2015). Analytical solution for bending analysis of functionally graded beam. Steel and Composite Structures, 19(14), 829–841.

Hwu, C., Hsu, H.W., & Lin, Y.H. (2017). Free vibration of composite sandwich plates and cylindrical shells. Composite Structures, DOI: http://dx.doi.org/10.1016/j.compstruct.2017.03.042.

Jane, K.C., & Harn, Y.C. (2000). Vibration of delaminated beam-plates with multiple delaminations under axial forces. Mechanical Structures and Machines, 28(1), 49–64.

Jun, L., Guangwei, R., & Jin, P. (2014). Free vibration analysis of a laminated shallow curved beam based on trigonometric shear deformation theory. Mechanics based Design of Structures and Machines, 42, 111–129.

Kant, T., & Swaminathan, K. (2001). Analytical solutions for free vibration of laminated composite and sandwich plates based on a higher-order refined theory. Composite Structures, 53, 73–85.

Karama, M., Afaq, K.S., & Mistou, S. (2003). Mechanical behaviour of laminated composite beam by the new multi layered laminated composite structures model with transverse shear stress continuity. International Journal of Solids Structures, 40, 1525–1546.

Khare, R.K., Kant, T., & Garg, A.K. (2004). Free vibration of composite and sandwich laminates with a higher-order facet shell element. Composite Structures, 65, 405–418.

Kulkarni, S.D., & Kapuria, S. (2008). Free vibration analysis of composite and sandwich plates using an improved discrete Kirchhoff quadrilateral element based on third order zigzag theory. Computational Mechanics, 42(6), 803–824.

Li, D.H., Wang, R.P., Qian, R.L., Liu, Y., & Qing, G.H. (2016). Static response and free vibration analysis of the composite sandwich structures with multilayer cores. International Journal of Mechanical Sciences, 111, 101–115.

Liu, B., Ferreira, A.J.M., Xing, Y.F., & Neves, A.M.A. (2016). Analysis of functionally graded sandwich and laminated shells using a layerwise theory and a differential quadrature finite element method. Composite Structures, 136, 546–553.

Loja, M.A.R., Barbosa, J.I., & Mota-Soares, C.M. (2015). Dynamic behaviour of soft core sandwich beam structures using kriging-based layerwise models. Composite Structures, 134, 883–894.

Lurllaro, L., Gherlone, M., Di-Sciuva, M., & Tessler, A. (2013). Assessment of Refined Zigzag Theory for bending, vibration and buckling of sandwich plates: a comparative study of different theories. Composite Structures, 106, 777–792.

Momeni, S., Zabihollah, A., & Behzad, M. (2017).Development of an accurate finite element model for N-layer MR-laminated beams using a layerwise theory. Mechanics of Advanced Materials and Structures, DOI: 10.1080/15376494.2017.1341579.

Natarajan, S., & Manickam, G. (2012). Bending and vibration of functionally graded material sandwich plates using an accurate theory. Finite Elements in Analysis and Design, 57, 32–42.

Nayak, A.K., & Satapathy, A.K. (2016). Stochastic damped free vibration analysis of composite sandwich plates. Procedia Engineering, 144, 1315–1324.

Neves, A.M.A., Ferreira, A.J.M., Carrera, E., Cinefra, M., Roque, C.M.C., Jorge, R.M.N., & Soares, C.M.M. (2013). Static, free vibration and buckling analysis of isotropic and sandwich functionally graded plates using a quasi-3D higher-order shear deformation theory and a meshless technique. Composites: Part B, 44, 657–674.

Phan-Dao, H.H., Thai, C.H., Lee, J., & Nguyen-Xuan, H. (2016). Analysis of laminated composite and sandwich plate structures using layerwise HSDT and improved meshfree radial point interpolation method. Aerospace Science and Technology, 15(44), 1–20.

Shariyat, M. (2010). A generalized global-local high-order theory for bending and vibration analysis of sandwich plates subjected to thermo-mechanical loads. International Journal of Mechanical Sciences, 52, 495–514.

Sharma, V.K., Kumar, V., & Joshi R.S. Effect of RE addition on wear behavior of an Al-6061 based hybrid composite. Wear. 2019 Apr 30;426:961–74.

Sharma, V.K., Kumar, V., & Joshi, R.S. Investigation of rare earth particulate on tribological and mechanical properties of Al-6061 alloy composites for aerospace application. Journal of Materials Research and Technology. 2019 Jul 1;8(4):3504–16.

Sharma, V.K., Kumar, V., & Joshi, R.S. Experimental investigation on effect of RE oxides addition on tribological and mechanical properties of Al-6063 based hybrid composites. Materials Research Express. 2019 Jun 5;6(8):0865d7.

Sharma, V.K., Aggarwal, D., Vinod, K., & Joshi, R.S. Influence of rare earth particulate on the mechanical & tribological properties of Al-6063/SiC hybrid composites. Particulate Science and Technology. 2021 Nov 17;39(8):928–43.

Stavski, Y. (1965). On the theory of symmetrically heterogeneous plate having the same thickness variation of the elastic moduli. Topics in Applied Mechanics American Elsevier, New York, 105.

Thai, H.T., & Choi, D.H. (2013). A simple first-order shear deformation theory for laminated composite plates. Composite Structures, 106, 754–763.

Versino, D., Gherlone, M., Mattone, M., Sciuva, M.D., &Tessler, A. (2013).C⁰ triangular elements based on the Refined Zigzag Theory for multilayered composite and sandwich plates. Composites: Part B, 44, 218–230.

Wang, M., Li, Z.M., & Qiao, P. (2018).Vibration analysis of sandwich plates with carbon nanotube-reinforced composite face-sheets. Composite Structures, 200, 799–809.

Chapter 6

Static and dynamic behavior analysis of Al-6063 alloy using modified Hopkinson bar

Ashish Bansal, Harsh Kumar Bhardwaj,
Vipin Kumar Sharma, and Ashwani Kumar

6.1 INTRODUCTION

Under static conditions, the stress intensity factor is proportional to the applied force. In the case of impact loading, the proportionality does not hold due to the effect of inertia. Bacon et al. (1994) emphasized that the effects of inertia have to be incorporated to evaluate the correct stress intensity factor. However, if the fundamental mode of vibration is predominant. The stress intensity factor is proportional to the load-point displacement. In this method, Bacon et al. (1994) evaluated the applied force and load-point displacement by a modified Hopkinson pressure bar, where two-point strain measurement was used. Bacon et al.(1994) used the load-point displacement and applied force at the end of the bar near the specimen to evaluate the

DOI: 10.1201/9781003360001-6

dynamic fracture toughness. The evaluation of load-point displacement and the applied force at the end of the Hopkinson bar was obtained in terms of strains and loads at the two different points on the Hopkinson bar. Bacon et al. (1994) also compared the method with the stress intensity factor derived from the strain measurement near the notch tip and good agreement between the two was obtained. This method is also well suited for high temperatures.

For measuring the dynamic fracture properties of the material at loading rates greater than 10^6 MPa $\sqrt{m}$ /s, a Hopkinson bar was modified by Jiang and Vecchio (2007). To better understand issues, "loss of contact" was first performed on a two-bar/three-point bend dynamic fracture test setup. To better understand the loss of contact phenomenon an electrical circuit was designed to record the output voltage. This electrical circuit could record the output voltage variations in interfaces. Based on this electrical circuit design, the output voltage of 0.75 V will be present when the circuit is complete. That is, the output voltage is 0.75 V when the impactor and specimen or specimen and supports are in contact with each other, whereas the circuit voltage will go to 1.5 V when loss of contact occurs. The effect of the pulse shaper on the duration of the incident pulse and dynamic stress equilibrium in the cracked three-point bend has been investigated.

Guo et al. (1997) used the finite element analysis method to analyze the stress behavior of simply supported a 40Cr Steel bar subjected to an impact load at the center of the backside of the test specimen. Rubio et al. (2003) is based on a three-point bending test at high loading rates for measuring the dynamic fracture toughness of the modified Hopkinson bar. T. Yokohama (1989) experimented on three-point bend specimens on Al 7075-T651 aluminum alloy to evaluate dynamic fracture initiation toughness. Three different procedures (input load, displacement, CMOD [crack mouth opening displacement]) were used to calculate the dynamic fracture initiation toughness. For the calculation of CMOD directly, a high-speed photography system was used. To validate the result obtained from three different procedures, numerical analysis was performed. After analysis, it was found that the result obtained from the numerical analysis is suited to the CMOD method. So for the SIF calculation, the CMOD method is the best way.

Yokoyama and Kishida (1989) used a split-Hopkinson pressure bar technique to measure the dynamic load applied to a three-point bend test specimen for determining the dynamic fracture initiation toughness. Bacon et al. (1994) studied the dynamic response of a notched three-point bend specimen. They used two-point strain measurement to measure the applied force and load-point displacement of a Hopkinson pressure bar near the notched tip. Gua et al. (1997) used the finite element analysis method to analyze the stress behavior of a simply supported 40Cr Steel bar subjected to an impact load at the center of the backside of the test specimen. Rubio et al. (2003) is based on a three-point bending test at high loading rates for measuring the dynamic fracture toughness of modified Hopkinson bar. Jiang and Vecchio (2007) modified a Hopkinson pressure bar to measure the dynamic fracture. Zing et al. (2013) investigated the dynamic fracture behavior of 2024-T4 and 7075-T6 aluminum alloy using an instrumental drop tower machine.

6.2 MATERIALS AND METHODS

6.2.1 Formula for compression experiment

$$\text{Strain rate: } \dot{\varepsilon} = \frac{2C_b\varepsilon_R}{h} \tag{6.1}$$

where C_b, ε_R, and h are the velocity of waves in the bar, reflected strain, and length of specimen, respectively.

$$\text{Strain in the specimen: } \varepsilon_s = 2\frac{C_b}{h}\int_0^t \varepsilon_R(t)\,dt \tag{6.2}$$

$$\text{Stress in specimen: } \sigma_s(t) = \frac{A_b}{A_s}E_b\varepsilon_T(t) \tag{6.3}$$

where A_b, A_s, E_b, and $\varepsilon_T(t)$ are the cross-sectional area of the bar, the cross-sectional area of the specimens, the Young modulus of the bar, and transmitted strain, respectively.

6.2.2 Split-Hopkinson pressure bar setup for compression

During dynamic experiments, the following assumptions were made:

- The wave propagates without any dispersion.
- The wave propagation is assumed to be one-dimensional.

Two bars are used in the split-Hopkinson pressure bar (SHPB) setup, one incident bar and the other a transmission bar (Figure 6.1). As the projectile comes and hits the incident bar, it generates a compressive stress wave in the incident bar. The strain measurement on the bars is performed with the help of strain gauges in the half Wheatstone bridge configuration. The specimen is placed between the two bars. To reduce the effects of interfacial friction, lubrication is used on the specimen surface and the bar ends. A momentum

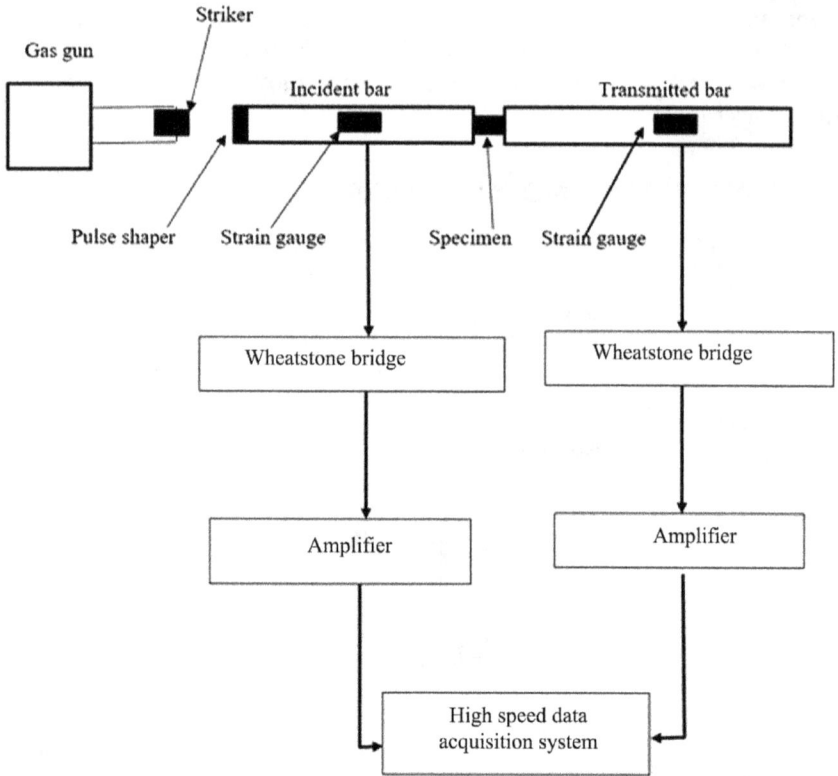

Figure 6.1 Schematic diagram of the split-Hopkinson pressure bar.

trap is also kept after the transmission bar to stop the transmission bar after impact.

A trapezoidal compressive stress wave is generated at the striking end of the incident bar by the striker. The specimen is placed between the incident and the transmission bar. The compressive wave propagates through the input bar and reaches the specimen. Some part of the incident pulse is reflected (ε_r) to the incident bar and the remaining part (ε_t) is transmitted to the transmission bar. The strain pulses are measured at the strain gauge stations on the bars. The obtained signal is shown in Figure 6.2. The speed of the striker and strain signals are recorded with a high-speed data acquisition system.

In dynamic loading such as Hopkinson bar experiments, the length of the sample is not a standard quantity and is decided so that it can fail under the equilibrium load on the contact surfaces with the bars. The use of specimens having a larger l/d (l = gauge length, d = diameter) ratio should be avoided

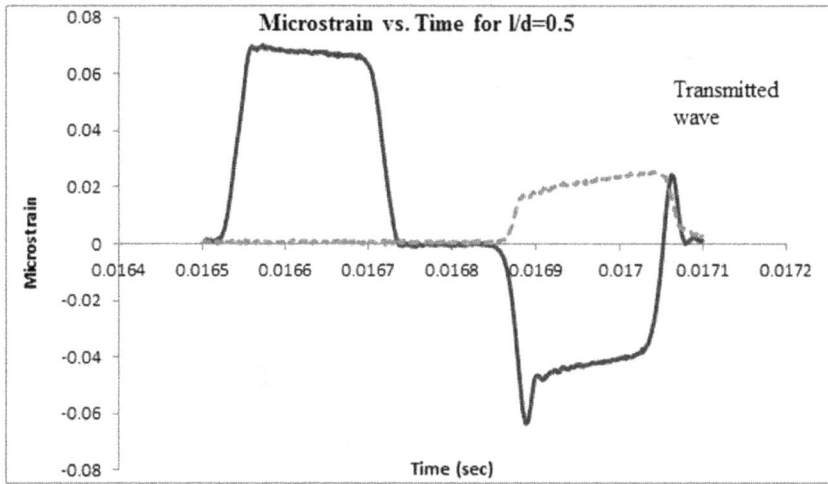

Figure 6.2 Signal obtained during SHPB experiments.

to prevent the buckling and shearing modes of deformation. The lower l/d ratio may increase friction on contact surfaces, which can raise the flow stresses. Therefore the optimized values of the l/d ratios are required to obtain the mechanical characteristics on better accuracy. Thus the cylindrical specimens of three different l/d ratios are prepared to understand the effects of l/d ratios in SHPB experiments.

6.2.3 Dynamic compression test for Al6063

For determining the mechanical properties of Al-6063, a compression test was performed on SHPB. The specimens of different length/diameter (l/d) ratios were prepared for the compression test. Here, three specimens of l/d ratio (0.5, 0.75, and 1.0) were made. During the compression test, the specimen was placed between the incident and transmission bars as shown in Figure 6.3.

The specimens before and after the experiment are shown in Figure 6.4. Three compression experiments were conducted on SHPB at the Impact Mechanics Lab of the Applied Mechanics Department.

The dimension before impact was length (fixed) = 5mm.

A schematic of the experimental setup of a modified SHPB, DAQ system, and ultra-high-speed cameras is shown in Figure 6.5. The compressor is used to fill the air in the cylinder. Instead of directly supplying the air from the compressor to the pressure chamber, the air is first filled in the cylinder

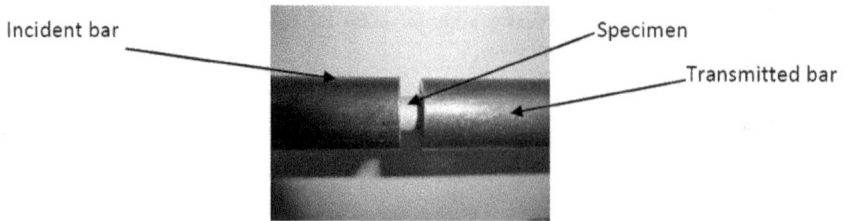

Figure 6.3 Specimen between incident and transmission bars.

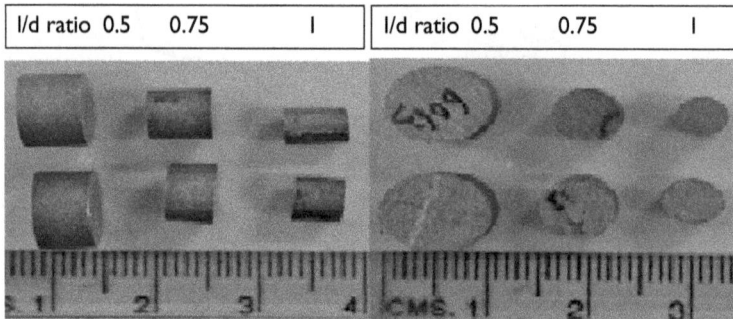

Figure 6.4 Specimen before and after impact along the length and diameter.

then it is supplied to the pressure chamber. When the air is directly supplied to the pressure chamber, it can contain moisture, which can cause corrosion. In-cylinder, the condensed air is moved down and the problem of corrosion does not come into play.

A barrel of length 2400 mm is used. At one end of the barrel, the pressure chamber is connected and at the end, the velocity probes are mounted. The velocity probes are used to measure the velocity of the projectile connected at the barrel end. The gap between the two ends of velocity probes is 50mm. The strain gauges are mounted on the Hopkinson bar. A specimen is placed between the end of the Hopkinson bar and the three-point bend fixture. The velocity probes and the strain gauges are connected to the data acquisition system. When a projectile is fired, it moves inside the barrel and impacts on the one end of the Hopkinson bar. A compressive pulse is generated, and it travels in the Hopkinson bar towards the interface of Hopkinson bar and specimen. The strain signals are recorded at the point of the Hopkinson bar where strain gauges are pasted. The strain signals are transferred to the computer through a data acquisition system.

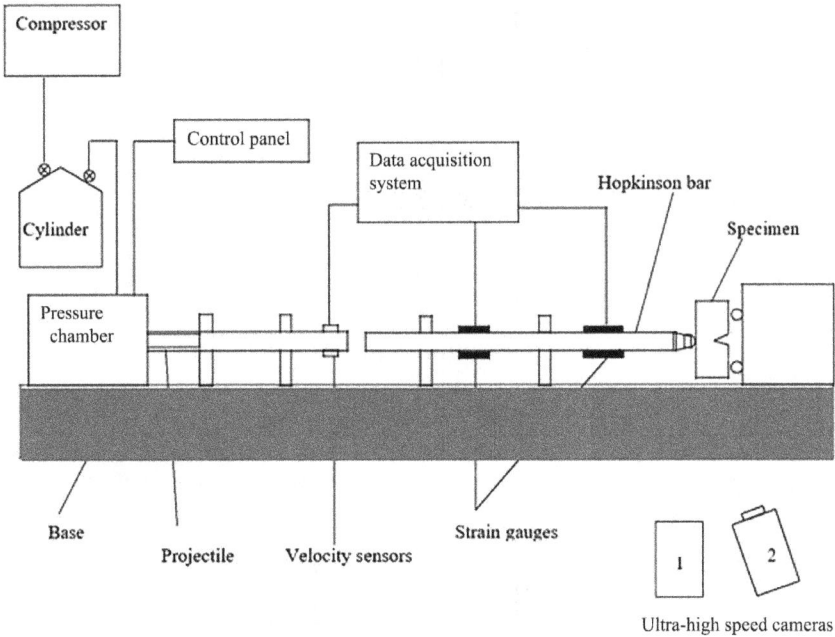

Figure 6.5 Schematic diagram of three-point bend setup on modified SHPB.

Table 6.1 Properties of Al- 6063 T6

Properties	Metric values
Density	2700 kg/m³
Modulus of elasticity	69 GPa
Yield strength	214 MPa

6.2.4 Spectroanalysis

The spectral analysis of Al 6063 T6 is conducted to determine the composition of the material. The properties of Al 6063 T6 are also shown in Table 6.1. The present alloy composition and standard alloy composition are shown in Table 6.2.

6.2.5 Specimen preparation

A specimen was machined according to E 399 standard, which is used for the three-point bend test as shown in Figure 6.6. The suitable dimension of

Table 6.2 Spectroanalysis of Al-6063 T6

Elements	Al	Si	Cu	Mg	Fe	Zn	Ni	Mn	Cr	V	Ti	Bi	Pb	Sn	Other each	Other Total
Present alloy composition (wt.%)	98.69	0.489	0.0176	0.5171	0.1834	0.0088	0.0016	0.051	0.0054	0.0037	0.0132	0.0016	0.0006	≤0.0010	–	–
Standard alloy composition (wt.%)	97.5	0.2–0.6	Max 0.1	0.4–0.9	Max 0.35	Max 0.1	–	Max 0.1	Max 0.1	–	Max 0.1	–	–	–	Max 0.05	Max 0.15

Figure 6.6 Three-point bend specimen (Al6063).

the specimen is 10mm ×20mm ×100mm. The dimensions of the specimen studied were thickness (10mm), width (20mm), and length (100mm).

The 10mm thick plate of material Al6063 was cut and finished in the required dimension by a circular cutter in the production lab. After that, a notch of uniform width 1.5mm and depth 6mm and further inclined width of depth 2mm, was made by use of two different circular cutters on a milling machine in the machining lab. A further fatigue notch of 2mm depth was made.

The HPB experiment setup, on which a three-point bend dynamic test will be conducted, is improved by changing the location of strain gauges at two places of the bar. Also, two photosensors, which were used for the velocity measurement of the striker bar, are replaced with new photosensors. Digital image correlation (DIC) can be used during the dynamic test.

6.3 RESULT AND DISCUSSION

6.3.1 Strain rate behavior of Al-6063 T6

The strain behavior of aluminum alloy, 6063 T6 were investigated using the three different specimens having the length 6 mm and diameters 6 mm, 8 mm, and 12 mm (l/d ratios of 1, 0.75, and 0.5). The dynamic compressive experiment was conducted without a pulse shaper in the split-Hopkinson pressure bar. For the velocity of projectile 9 m/s, the strain rate obtained for the specimens is shown in Figure 6.7.

The variation of strain rates with the times of three different l/d ratios is shown in Figure 6.7. It is concluded from the figure that when l/d ratios increase the strain rates decreases because, according to the preceding formula, the strain rate is inversely proportional to the length of the specimen. So the strain rate is the largest for the specimen with an l/d ratio equal to 0.5, and the strain rate is lowest for specimen with an l/d ratio of 1 (Figure 6.7). The strain rates increase to 1350 s^{-1}, 1600 s^{-1}, and 3550 s^{-1}, when the pulse shaper is not used in the experiments.

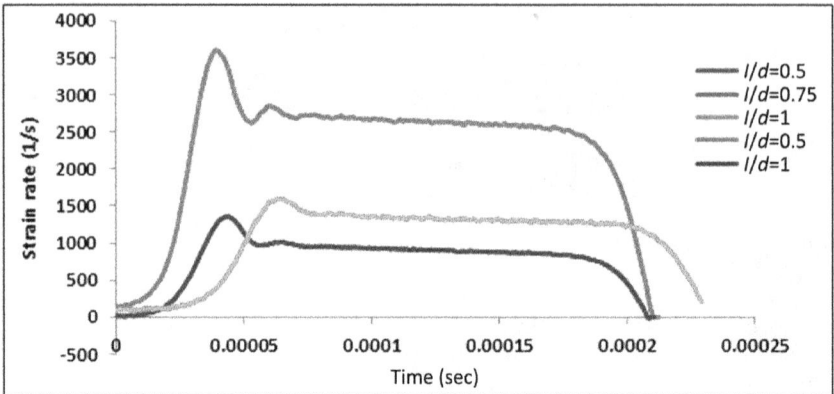

Figure 6.7 Strain rate vs. time.

For the *l/d* ratio of 0.5, it is observed by a graph that for the time up to 45μs, the strain rate increases up to 3600 s⁻¹; after that, its value decreases from 3600 to 2800 s⁻¹. And after 50μs, it becomes almost constant.

The strain rate increases due to increases in dynamic load on the specimen, and before the specimen obtains an equilibrium state, some oscillation occurs in the specimen. When equilibrium is attained, then the strain rate becomes almost constant. And finally, after 180μs the strain rate decreases due to the unloading of the specimen.

For *l/d*=0.75, it follows a similar curve, the strain rate increases sharply up to 1600 s⁻¹ and then decreases to 1380 s⁻¹ up to time 70μs. After that, the strain becomes almost constant to 210μs. The strain rate for *l/d*=1 follows the same pattern and increases up to 1350 s⁻¹.

Figure 6.8 shows the variation of engineering stress with engineering strain for *l/d*=0.5. The maximum stress obtained in the specimen is approximately 650MPa. For the *l/d* ratio of 0.75, the maximum stress obtained in this specimen is 380MPa. For *l/d*=1, the maximum stress obtained in this graph is 230MPa.

The stress-vs.-strain graph of different *l/d* is plotted without using the pulse shaper. The strain hardening in the material of *l/d*=1.0 is high, as shown in Figure 6.8.

6.3.2 Quasi-static test on three-point bend specimen

A quasi-static test was conducted on the notched three-point bend specimen (E399 standard) of the Al6063 T6 at the constant crosshead speed of 0.5mm/min. From the extracted data of load and crack mouth opening

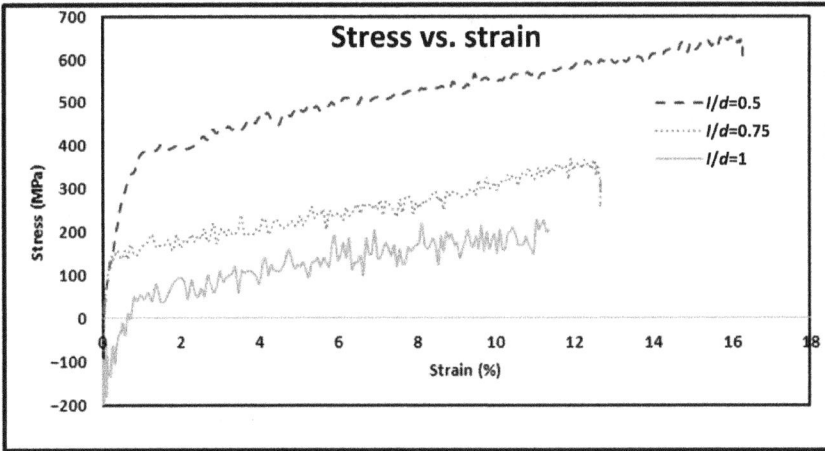

Figure 6.8 Stress vs. strain for l/d = 0.5, 0.75, and 1.

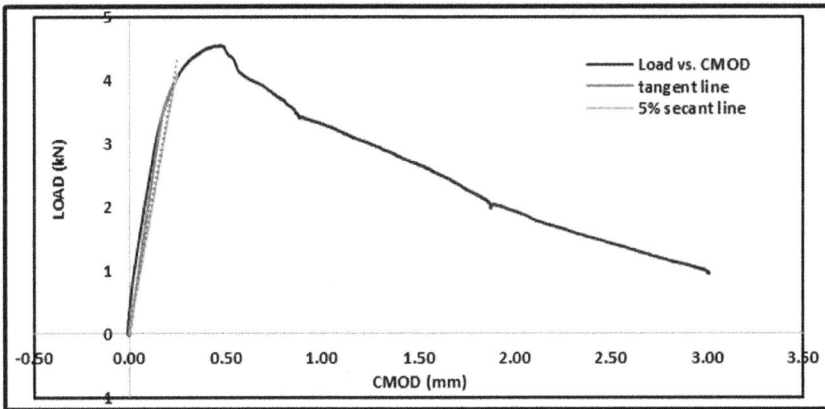

Figure 6.9 Load COD curve for quasi-static test on a notched specimen.

displacement (CMOD), plot load vs. CMOD is drawn. It was observed that the first load increases with deflection, reaches a maximum value, and then decreases again, as shown in Figure 6.9. This decrease in the load may be due to crack propagation in the notched specimen.

For calculation of K_Q from the static fracture toughness formula, the load is substituted from the load-vs.-CMOD curve. This load is obtained by the intersection of the original load-vs.-CMOD and the second curve, which is a straight line and has a slope of 95% of the elastic portion of the original curve. The plane-strain fracture toughness K_{IC} is calculated for the three-point bend specimen by using the standard formula as

$$\mathbf{K_Q} = \frac{3FS\sqrt{a}}{2BW^2} Y\left(\frac{a}{W}\right) \tag{6.11}$$

where K_Q is the tentative fracture toughness, and F is the applied load at the fracture initiation. The plane-strain criteria for the valid K_{IC} value are given by

$$B \text{ and } a \geq 2.5 \ (K_Q/\sigma_Y)^2 \tag{6.12}$$

where σ_Y is the 0.2% offset yield strength of the material. The value of B obtained from equation (6.12) is 9.5mm, and the value considered for specimen thickness in static condition test was 10mm. So the condition for the plane-strain requirement is satisfied. K_Q can be written as K_{IC}. The K_{IC} value for the 6063 T6 aluminum alloy is 13.5 MPa√m. The K_{IC} value obtained for the material is at room temperature.

6.3.3 Dynamic test on three-point bend specimen

Dynamic test on the modified SHPB was also conducted on the notched specimen on Al6063 T6 material.

Data obtained from the strain gauges acquired by the data acquisition system were in the form of volts, which were later converted into strain using the Wheatstone bridge parameters and using the modified SHPB theory for the three-point bend test. Load-point displacement and normal force data at various positions were obtained. For plotting the curves, MATLAB was used. The experiment was conducted, and graphs were plotted in this section for test conditions at 1 bar, 1.5 bar, and 2 bar.

Experiment I

A three-point bend dynamic test is conducted on 1 bar pressure, and the projectile is impacted on the input bar with a velocity of 9.14 m/s. Figure 6.10 shows the variation of forces with time. Na indicates the force at the Hopkinson bar at point A. Nb indicates the force at the Hopkinson bar at point B. Ne indicates the force at the end of the Hopkinson bar. The maximum force at point E is 58kN at 175µs. The loading duration and maximum value of force obtained from strain gauges posted at point A on the Hopkinson pressure bar are 120µs and 50kN, respectively. Similarly, the loading duration and maximum value of force obtained from strain gauges posted at point B on the Hopkinson pressure bar are 120µs and 45kN, respectively. The load at E is obtained by the use of load at point A and particle velocity at point A. The load at point E is obtained through MATLAB. The maximum value of force and duration at point E is 60kN and 60μs.

Figure 6.11 shows the variation of displacement with time. The maximum displacement obtained is 1.1 mm. First, the displacement increases slowly up to 175μs, then it increases sharply up to 300μs, and then finally becomes almost constant from 300μs onward.

Na, Nb, and Ne indicate the load-point force at points A, B, and C, respectively, in Figure 6.10. The load-point displacement at point E is shown in Figure 6.11, and it indicates that the load-point displacement increases with time.

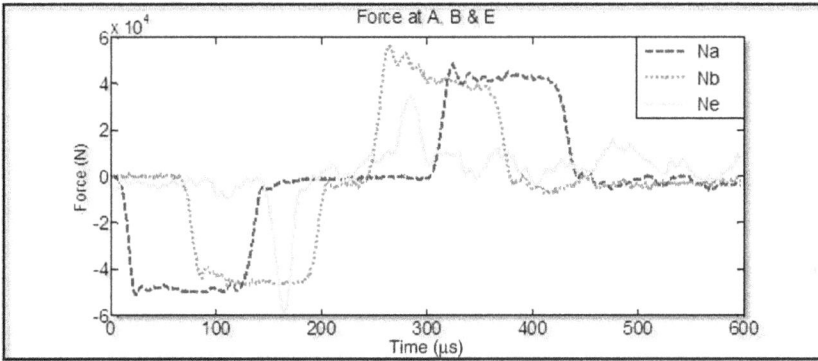

Figure 6.10 Normal force variations with time curve on the notched specimen at 1 bar.

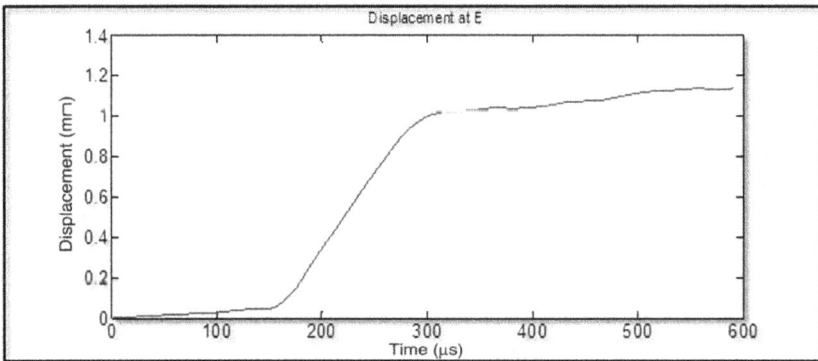

Figure 6.11 Load-point displacement curve on the notched specimen at 1 bar.

Experiment 2

The test is conducted on 1.5 bar pressure and projectile of velocity 12 m/s. Figure 6.12 shows the variation of forces with time. Na indicates the force at the Hopkinson bar at point A. Nb indicates the force at the Hopkinson bar at point B. Ne indicates the force at the end of the Hopkinson bar. Figure 6.13 shows the variations of displacement with time.

According to Figure 6.12, the maximum force at point E is obtained by the use of force and velocity at point A. The maximum value of force at point E is approximately 70 kN, which is higher than the value obtained at 1 bar pressure. Figure 6.12 shows that the force at point E is maximum at time 200μs.

The maximum displacement obtained at the load-point is 2.1mm (Figure 6.13). The displacement at point E is calculated by integrating the

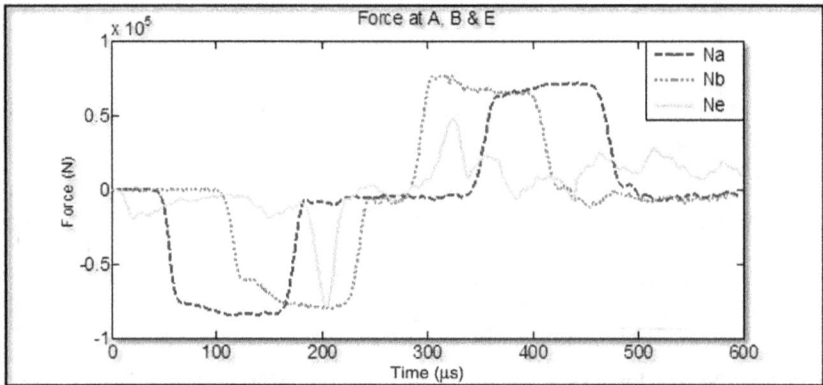

Figure 6.12 Normal force variations with time curve on the notched specimen at 1.5 bar.

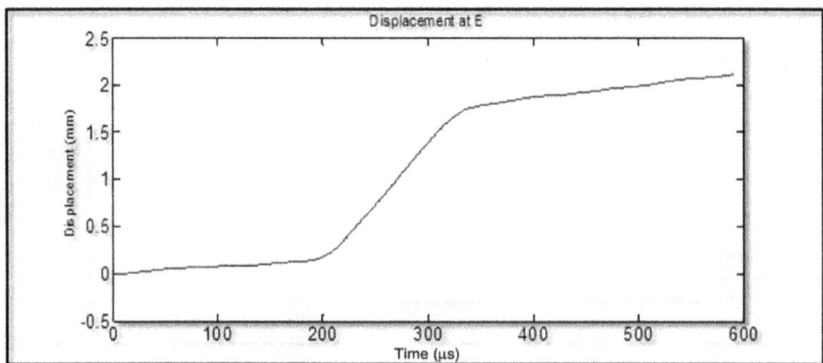

Figure 6.13 Load-point displacement curve on the notched specimen at 1.5 bar.

velocity at point E. In Figure 6.13, the displacement starts increasing slowly up to 200 μs. During 200 to 300μs, it increases sharply and after that again increases slowly.

Experiment 3

The test was conducted on 2 bar pressure on a three-point bend specimen. The striker bar velocity measured at the end of the barrel before impact to the Hopkinson bar was 14 m/s. From strain measured at points A and B, the value of force and displacement is obtained at point E. The variations of force at points A, B, and E with time are shown in Figure 6.14. From that figure, it is observed that the nature of the variation of load-point force and load-point displacement is similar to pressure at 1 bar and 1.5 bar. The maximum force at point 'E' is 80kN.

The maximum displacement at point E is also obtained by integrating the velocity at point E similarly to the previous experiment conducted at 1 bar and 1.5 bar. The trend of the displacement with time is shown in Figure 6.15. It is observed that the displacement increases from time 240 to 370μs and that it attains a maximum value of 2.05 mm after 370μs.

From these experiments conducted at 1 bar, 1.5 bar, and 2 bar, it is observed that when pressure increases, then load-point force and load-point displacement increase. The increase in load and load-point displacement during varying pressure from 1 to 1.5 bar is more in comparison to the change in pressure from 1.5 to 2 bar. From Figures 6.11, 6.13, and 6.15, which are drawn between displacement vs. time, it is observed that, initially, displacement is almost constant and that it then rises sharply and finally becomes constant. The duration of the sharp increment of displacement for 1, 1.5, and 2 bar are 160, 140, and120μs, respectively. So it is concluded

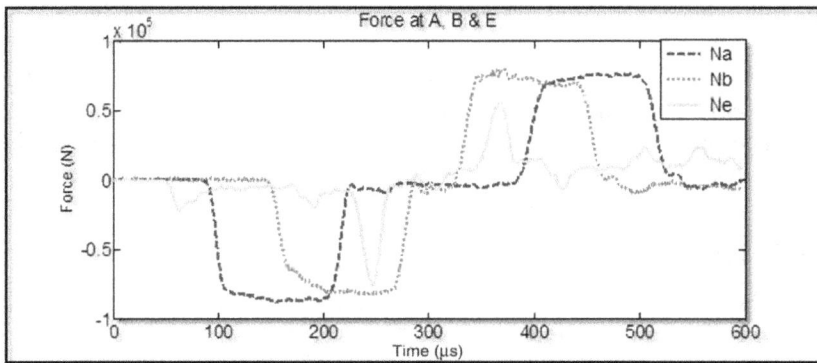

Figure 6.14 Normal force variations with time curve on the notched specimen at 2 bar.

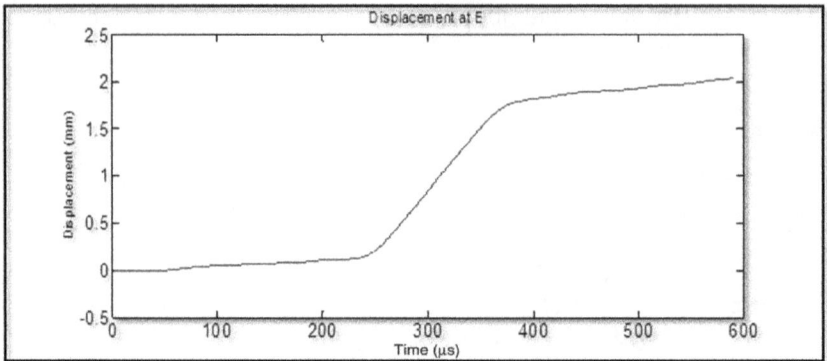

Figure 6.15 Load-point displacement curve on the notched specimen at 2 bar.

that when pressure rises, then the duration of the sharp rise of displacement decreases. The decrease in the duration of time occurs due to increases in the velocity of the striker.

6.4 CONCLUSIONS

6.4.1 Quasi-static test

- It is found that first the load increases with deflection, reaches a maximum value, and then decreases again.
- This decrease in the load is due to crack propagation in the notched specimen.

6.4.2 Dynamic test

The following conclusions are derived from the experimental analysis:

- As the velocity of impact increases, the normal force at the point E increases, and the displacement also increases.
- The duration of the sharp increase of displacement decreases as the velocity of the striker increases.

6.4.3 Strain rate analysis

- The strain rate increases with the decrease in the length-to-diameter (l/d) ratio of the specimens.
- The engineering stress in the specimen is highest when the l/d ratio is minimum, and it is lowest when the l/d ratio is maximum.

REFERENCES

Bacon, C., Farm, J., & Lataillade, J.L. (1994).Dynamic fracture toughness determined from load-point displacement. *Experimental Mechanics*, 34(3), 217–223.

Guo, W.G., Li, Y.L., & Liu, Y.Y. (1997). Analytical and experimental determination of dynamic impact stress intensity factor of 40Cr steel. *Theoretical and Applied Fracture Mechanics,* 26(1), 29–34.

Jiang, F., Vecchio, & K.S. (2007). Experimental investigation of dynamic effects in a two bar/three point bend fracture test. *Review of Scientific Instruments*, 78(6), 063903.

Rubio, L., Fernandez-Saez, J., & Navarro, C. (2003). Determination of dynamic fracture initiation toughness using three point bending tests in a modified Hopkinson pressure bar. *Experimental Mechanics,* 43(4), 379–386.

Xing, M.Z., Wang, Y.G., & Jiang, Z.X. (2013). Dynamic fracture behaviours of selected aluminum alloys under three point bending. *Defence Technology*, 9(4), 193–200.

Yokoyama, T., & Kishida, K. (1989). A novel impact three point bend test method for determining dynamic fracture initiation toughness. *Experimental Mechanics*, 29(2), 188–194.

Chapter 7

Thermal stress analysis and parametric study of thermal barrier coated engine piston

Subodh Kumar Sharma, K. V. Ojha, and Shekhar Bhardwaj

7.1 INTRODUCTION

The components of a diesel engine combustion chamber, like valves, pistons, piston heads, and cylinders, work at elevated temperatures, thus increasing the chances of their failure. Therefore, thermal barrier coatings (TBC) application has been considered to play an essential role in the improvement in the lives of these components. These coatings permit higher performing temperatures while restraining heat contact of structural elements and improving component life by lowering thermal fatigue and oxidation. Interest in the production of low-weight, thermal-resistant pistons in combustion engines has strongly intensified due to the economic and ecological aspects of car engine operation. However, beyond a certain thickness, TBC acts as a rigid plate and is further associated with the following issues: it increases parasitic weight on the part, which affects the rotating components; it increases residual stress and strain, which causes coating failure. It increases the possibility of sintering and associated structural and thermal problems due to its own increased surface temperature.

TBCs are well developed material systems generally introduced to metallic surfaces, such as on pistons, valves, gas turbines, or aero-engine parts, etc., working at higher temperatures. Depending upon the specific application and the operating conditions involved, the thickness of a typical TBC may vary from 100 micro-m to 2.0 mm. By virtue of its low thermal conductivity,

DOI: 10.1201/9781003360001-7

TBC provides insulation to the components from high heat loads as its material is capable of maintaining large temperature gradients between the coating surface and load-bearing alloys by virtue of its low thermal conductivity. Therefore, these coatings permit higher functioning temperatures while restraining the thermal contact of structural elements and improving component life by reducing thermal fatigue. Owing to the growing demand for higher efficiency at higher temperatures, better strength, and durability, the development of new and advanced TBCs has become a necessity.

A typical TBC consists of three layers (schematically described in Figure 7.1): the ceramic topcoat, metallic bond coat, and metal substrate. In general, the ceramic topcoat is composed of yttria-stabilized zirconia (YSZ), which is well-known for having extremely low conductivity, a high melting point, and high temperature stability. This ceramic layer (YSZ) maintains the thermal gradient of the TBC top layer and keeps the substrate at a lower temperature. The adhesion layer, known as bond coat layer, is an oxidation-resistant metal layer that is applied onto the metal substrate. Generally, it is 80 - 45 micro-m thick and made of NiCrAl alloy. Its primary function is to protect the metal substrate especially from corrosion elements and oxygen that permeate through the porous ceramic TBC. Other advantages of TBC include shielding and protecting the combustion chamber components from thermal fatigue and stresses, as well as lowering cooling requirements. TBCs were successfully tested for the first time in the gas turbine engine in the mid-1970s [1]. The TBC used therein was a duplex material system intended to fulfill the objective of shielding gas turbine parts from intense thermal environments and thus to enhance performance while reducing unnecessary emissions.

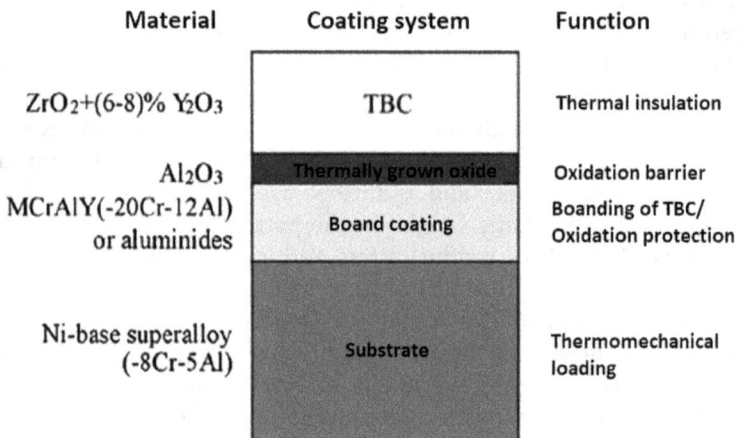

Material	Coating system	Function
ZrO_2+(6-8)% Y_2O_3	TBC	Thermal insulation
Al_2O_3	Thermally grown oxide	Oxidation barrier
MCrAlY(-20Cr-12Al) or aluminides	Boand coating	Boanding of TBC/ Oxidation protection
Ni-base superalloy (-8Cr-5Al)	Substrate	Thermomechanical loading

Figure 7.1 Typical thermal barrier coating system.

The ability of TBC to improve fuel economy depends on various factors, including the combustion chamber design, fuel injection system, and engine fuel economy. Several researchers [1–4] have worked on the measurement of heat transfer effects on the IC (internal combustion) engine components, mainly the valve, piston, and cylinder head, and they found a good correlation between the temperature distribution in these components and the heat produced via the combustion of fuel. Many researchers [5-7] have worked to measure valve and piston temperatures during operation and found their dependence on spark timing, engine cooling, and air–fuel mixture ratios.

The thermal conductivity (K) of TBC materials varies between 0.5 and 10W/mK, and the TBC thickness lies between 0.1 and 2.5 mm [8–10]. Several techniques are available for the deposition of ceramic coatings on metallic substrates; however, plasma spray coatings are used most expensively as TBCs [11]. Ceramics have low coefficients of thermal expansion (CTEs); the approximate range of CTEs varies from $3 \times 10^{-6}/°C$ for SN to $10 \times 10^{-6}/°C$ for YSZ, compared to $24 \times 10^{-6}/°C$ for aluminum and $12 \times 10^{-6}/°C$ for steel and iron [12]. Due to this difference in CTEs, thermal stresses differ at the interfacing layer between the ceramic coating and the metal substrate. This stress is predominantly large when substrate material is aluminum. As ceramic is brittle in nature, small imperfections can easily propagate into large cracks and the coating may separate from the bonding layer [13].

Several attempts have been made over the last four decades by researchers across the globe to prevent heat loss and improve the overall efficiency of the engine. A layer of 4.5 mm SN disc was placed over an air gap on the piston crown to raise the surface temperature by several hundred degrees. It was found that heat flux into the insulated piston crown surface was maximum at the time of the expansion stroke as compared to an aluminum alloy piston. Similar investigations were conducted during experiments on a thermal barrier coated piston [14–16]. These studies found that due to changes in the convective heat transfer coefficient, thermal boundary layers get heated at appreciably higher surface temperatures. Shalev et al. [17] explored theoretically and also carried out experimental investigation for crack development in the cylinder heads, and they noted that the cracking phenomenon was related to the low-cycle thermal fatigue mechanism induced by engine start-up, loading, and shutdown. One of the most noteworthy findings was related to increased hot-surface temperature. Even at a temperature of 590 °C, as compared to 350 °C, it was also quite safe as no cracks were noted at this elevated temperature. This effect was attributed to the proper functioning of the cooling system and its response to increased thermal loading. As a remark, Shalev et al. [17] reported that thermal-stress-related engine failure can be prevented by enhancing local cooling in the critical zones.

Analysis of temperature distributions in a real-time operational engine piston has been demonstrated [18] using FEM (finite element method) heat transfer models. For the purpose, the time-mean, area-averaged gas temperatures were taken in addition to the heat transfer coefficient using engine simulation. In the same manner, Ong [19] documented the results of a FEA-based forecast of the steady state temperature distribution in a diesel engine piston. In this work, MSC/NASTRAN, was employed for the investigation. A good correlation was observed between predicted and measured temperatures. Thermal analysis of an IC engine piston was conducted, and numerical calculations based on FDM were carried out to assess the transient thermal response of an aluminum alloy piston having 1 cm thick cylinder wall [20]. Several researchers have been working on advanced TBCs to improve the performance of engine components. Reduction in fuel consumption at low compression ratios after coating of a piston crown with 0.15 mm of zirconium was observed [21]. Nakic et al. [22] did experimental work to analyze the effect of surface temperatures on deposit growth. Thermal barrier coating is also recommended as there are opportunities for emissions improvements. TBCs application in two-stroke SI (spark ignition) engines was discovered to provide an ecofriendly environment by reducing carbon monoxide and hydrocarbon emissions [23, 24], owing to the complete consumption of fuel.

Based on the TBC investigation, Buyukkaya and Cerit [25] conducted thermal analyses of a conventional diesel piston made of aluminum silicon alloy and steel. Thermal analysis was executed on pistons coated with MgO material by using ANSYS. A computational analysis was performed to estimate the thermal stresses in the combustion chamber of an SI engine. Static thermal analysis was performed to determine the temperature variations in pistons coated with 1 mm thick plasma-sprayed yttria-stabilized zirconia [26, 27]. Numerous efforts have been made by the researchers to explore the effect of TBC thickness on temperature, and it has been concluded that the temperature drop is proportional to the thickness of the coating. So the study further continued to find the effect of TBC thickness, which is applied to the head of the exhaust valve and piston. Here, analyses of temperature, thermal distortion, and thermal stress have been carried out using different thicknesses of TBC, varying from 0.2 to 1.0 mm.

7.2 MATERIALS AND METHODS

The problem considered here is to investigate the temperature and distortion in the piston of a diesel engine, for specified barrier coatings (BCs). On the basis of the available temperature field, it is proposed to determine thermal stresses in the piston body. It is further proposed to investigate the effects of YSZ-TBC on piston crowns with different thickness values and compare the output characteristics with those pertaining to conventional pistons made of aluminum alloy.

Among all the methods, the variational method has proved itself as the first choice of researchers for the formulation of element relationships in FEA. The aforesaid method includes the selection of the proper variational principle followed by the expression of the function involved in terms of approximate assumed displacements, to comply with the stated boundary conditions. A set of governing equations is then developed by minimizing the approximate functional. For the heat transfer modes like conduction and convection and for contact boundary surfaces, variational formulations are derived.

The investigation of heat transfer can be carried out by various methods, like the heat electrical analogy method, the finite difference method, and the FEM. In the present study, a FEM with triangular elements has been applied to determine the temperature field in the piston components of an internal combustion diesel engine. The temperature field, so obtained, is used in further investigations into thermal stresses. For the solution of the problem, the piston body is divided into triangular elements, and the coordinates for each node are determined. The necessary boundary conditions are then imposed on the piston body.

As reported earlier, the formulations of finite element analysis need to start from the development of variational statements of the problem following suitable shape functions. By using suitable shape functions, a set of algebraic equations is developed. The necessary boundary conditions imposed on the piston body pertain to convective heat transfer on the three sides, including water, air, and the combustion gas sides of the piston. In addition to convective heat transfer, the necessary BCs are also imposed on heat transfer through the contact boundary between the piston and piston rings. Furthermore, the elemental equations developed by the minimization of variational integrals are assembled, and a set of simultaneous equations are obtained. Those equations are solved to achieve the nodal values of the field variables under consideration.

The heat balance approach was used in this study to validate the heat transfer model. For steady state conditions, as per the principle involved in the conversion of energy, the heat exchange equilibrium is maintained by the constant heat supply to the piston head from combustion gas and its release to the surroundings through air/water. Present analysis regarding the temperatures at all the nodal points available in the models reveals the accuracy as they completely comply with the heat flow BCs.

The piston chosen for this study has a diameter of 80 mm and a height of 105 mm for the objective. In the present study, the core material of the piston includes aluminum, which is further coated with YSZ for thermal insulation, especially on its head portion. For the fixation of YSZ on piston heads, nickel-chromium alloy is used as a bonding material. However, the

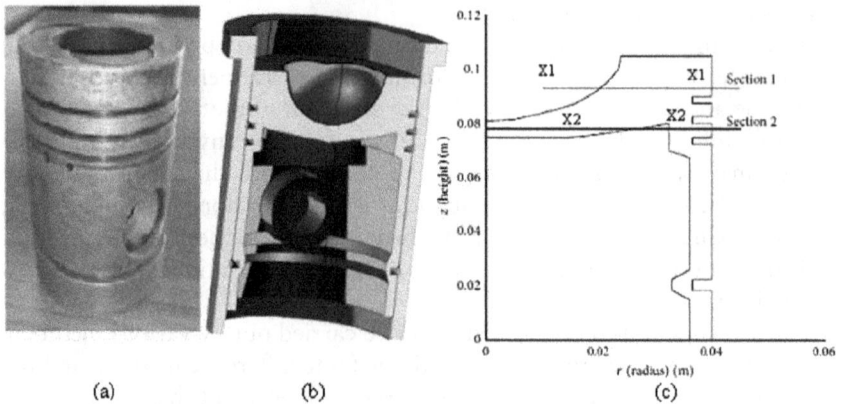

Figure 7.2 Piston (a) actual view, (b) cut section view, and (c) axis symmetric model.

piston ring and the cylinder wall are made of steel and cast iron, respectively. Figure 7.2 represents the piston of the chosen engine, which shows the sections X_1–X_1 and X_2–X_2 chosen as the two representative cross sections at which temperature, thermal distortion, and thermal stresses are plotted. For easy understanding, the mesh half-cross section of the piston has been shown in Figure 7.3 along with their boundary conditions. Figure 7.3 also represents the axis-symmetric meshed piston model, which depicts the discretization of the piston, piston rings, and cylinder wall into 395 triangular elements with a total of 311 nodes. The convective heat transfer coefficient and hot combustion gas temperature at piston top surface are Hg and Tg, respectively. The combustion gas temperature and heat transfer coefficients are assumed to be constant as it is the case of steady state condition. To determine these constant values, the instantaneous value of heat transfer coefficient (Hg) on the gaseous face at any crank angle (ϕ) is obtained using the formula given by Eichelberg [2]. As we know there is a great change in the temperatures of cooling water (20°C) and air (25°C) at the starting phase, which at steady state increased to 80°C and 85°C respectively. Therefore, the constant temperatures, i.e. 80°C and 85°C of water and air, respectively, have been taken into account. The boundary conditions, which are applied to the piston body, are indicated in Table 7.1 [28]. Based on these boundary conditions, temperature, deformation, and thermal stresses were analyzed.

7.3 RESULTS AND DISCUSSION

7.3.1 Impact of TBC on piston temperature

The computational results pertaining to the piston are depicted in Figures 7.4–7.11. Figure 7.4 represents the temperature distribution in the conventional

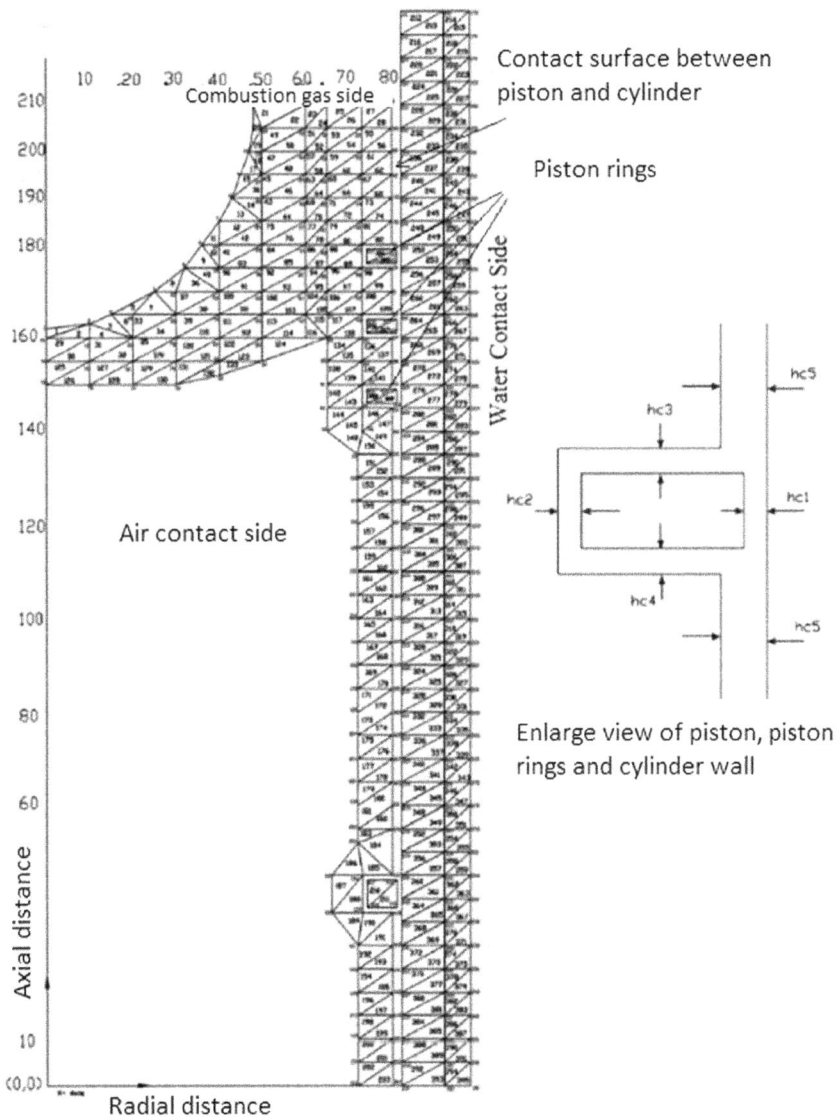

Figure 7.3 Axis symmetric meshed piston model (scale 2:1).

piston at sections X_1–X_1 and X_2–X_2. Similarly, for TBC- coated pistons with varying coating thicknesses, the temperature distributions are shown in Figure 7.5. In both Figures 7.4 and 7.5, it can be seen that a maximum temperature is found at the top of the piston. This finding in the current study is supported by the fact that the nodal point in the middle of the head

Table 7.1 Heat transfer parameters for piston body [28]

Piston	Temperature (°C)	T_g (Comb. side)	1000	Ta (Air side)	85
		T_w (Water side)	80		
	Coefficient of heat transfer (W/ m²K)	H_g (Comb. side)	290.5	hc_1	38346.0
		H_w (Water side)	1859.2	hc_2	20.0
		H_a (Air side)	174.3	$hc_3 = hc_4$	290.5
				hc_5	2324.0

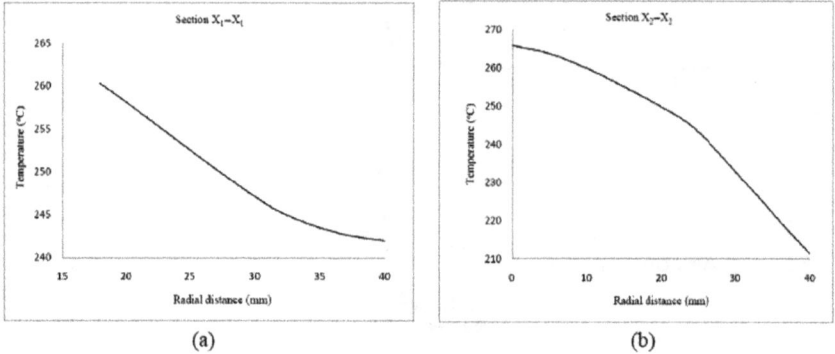

(a)   (b)

Figure 7.4 Temperature distribution in conventional piston body at (a) section $X_1- X_1$ and (b) section X_2-X_2.

has the highest temperature when subjected to the combustion gases. In contrast, the minimum temperature has been observed at the piston bottom surface that is open to air. In conventional pistons and for the entire range of coating thickness, it was observed that the temperature gap between the center of the piston head and its periphery is 25 °C at section X_1-X_1 and 45 °C at section X_2-X_2. These findings directly suggest the involvement of a "temperature difference" as a major factor for crack initiations in the piston by inducing thermal stresses.

For conventional pistons (Figure 7.4), the maximum temperature is found to be 260.29°C at section X1–X1 and 265.65°C at section X_2-X_2. This higher surface temperature leads to structural and thermal problems. By using TBC, there is an assurance that the life span of the piston will be significantly enhanced because this ceramic layer creates a very high thermal gradient and allows the lower layers to remain at a relatively lower temperature. Figure 7.5 gives a clear picture that the corresponding temperatures for the thermal barrier coated pistons are lower than the conventional piston. For 1 mm thick coatings, the maximum temperature drops of 23 °C and 25 °C were observed at X_1-X_1 and X_2-X_2, respectively. As previously concluded, the effectiveness of TBC increases with increasing coating

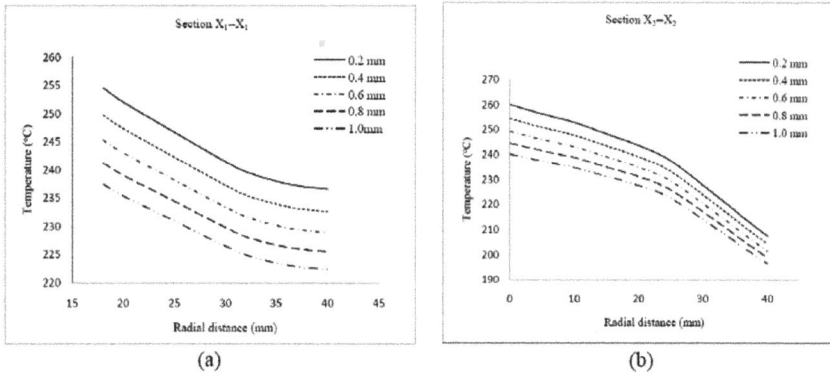

Figure 7.5 Temperature distribution in TBC piston body with 0.2, 0.4, 0.6, 0.8, and 1.0 mm coating thickness at (a) section X_1–X_1 and (b) section X_2–X_2.

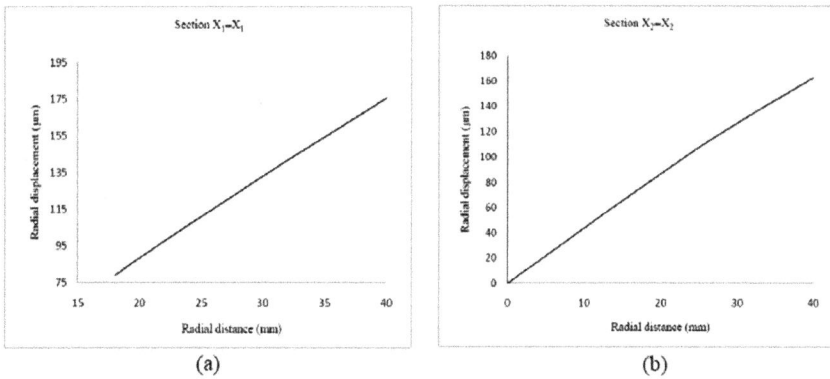

Figure 7.6 Radial displacement in conventional piston body at (a) section X_1–X_1 and (b) section X_2–X_2.

thickness. During the investigation, it was found that the temperature of the conventional piston was found to be higher (2.2–8.8%) than the thermal barrier coated piston at section X_1–X_1 and 2.16–9.58% at section X_2–X_2.

7.3.2 Impact of TBC on distortion in piston

Figure 7.6 shows the radial displacement of the nodal points, which are on sections X_1–X_1 and X_2–X_2 in conventional pistons, while Figure 7.7 shows the same for thermal barrier coated pistons. In both cases, the radial displacement at the symmetrical axis is nearly zero and the maximum at the outer periphery.

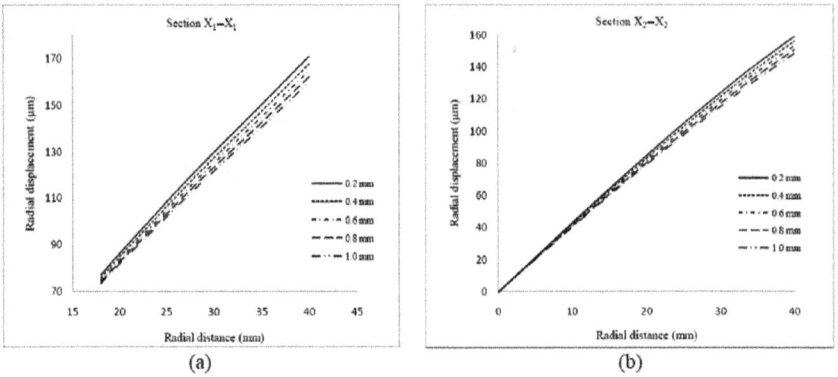

Figure 7.7 Radial displacement in TBC piston body with 0.2, 0.4, 0.6, 0.8, and 1.0 mm coating thickness at (a) section X_1–X_1 and (b) section X_2–X_2.

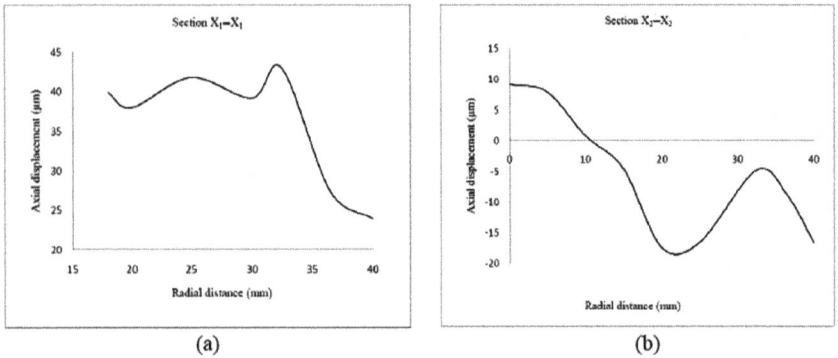

Figure 7.8 Axial displacement in conventional piston body at (a) section X_1–X_1 and (b) section X_2–X_2.

From Figure 7.6, the maximum radial displacements observed were 0.175 mm and 0.159 mm at sections X_1–X_1 and X_2–X_2, respectively. From Figure 7.7, the corresponding radial displacements for the case of thermal barrier coated pistons are lower than the previous values. The variation of radial displacement with respect to coating thickness is very small, and it increases linearly from the center to the outer periphery of the piston. Radial displacement slightly decreases as TBC thickness increases. Reduction in radial displacement attributed to TBC ranges from 2.56 to 8.89% at section X_1–X_1 and from 2.45 to 8.96% at section X_2–X_2 for the range of coating thickness considered here.

Similarly, Figures 7.8 and 7.9 represent the axial dislocation of nodal points for conventional and thermal barrier coated pistons, respectively.

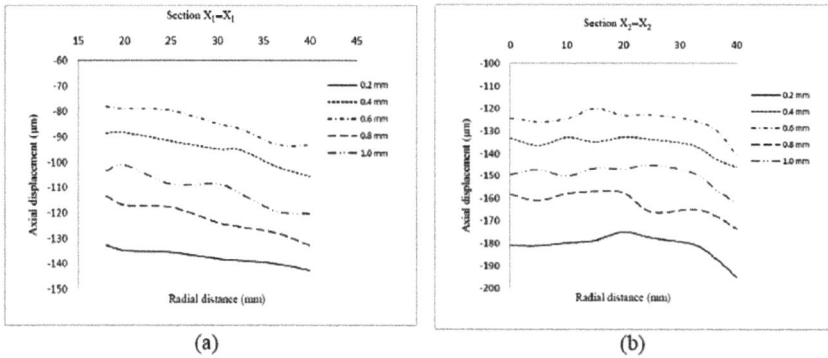

Figure 7.9 Axial displacement in TBC piston body with 0.2, 0.4, 0.6, 0.8, and 1.0 mm coating thickness at (a) section X_1–X_1 and (b) section X_2–X_2.

These figures indicate a large variation in the magnitude and direction, which is due to the fact that there is a tendency for each element to maintain a constant strain.

For conventional pistons, maximum axial displacements of 0.043 and 0.017 mm are observed at sections X_1–X_1 and X_2–X_2, respectively. Figure 7.9 clearly shows that the corresponding values for the TBC piston are much higher than for the conventional piston. The corresponding values for TBC pistons were found to be 0.120 and 0.162 mm for a coating thickness of 1.0 mm. The maximum and minimum values of axial displacement are observed for coatings of 0.2 and 0.6 mm, respectively.

7.3.3 Impact of TBC on thermal stresses in piston

Based on specified boundary conditions, a maximum von mises stress of 234.55 MPa is observed at section X_1–X_1 and 192.74 MPa at section X_2–X_2 in conventional pistons in Figure 7.10. As seen from Figure 7.11, after coating the piston, the maximum value of von mises stress drops down to 220.82 and 190.84 MPa at sections X_1–X_1 and X_2–X_2 for 1.0 mm coating thickness, respectively. Further, there is clear evidence that minimal variation in von mises stress is observed at 0.6 mm coating thickness.

7.4 VALIDATION OF PISTON MODEL

To confirm the heat transfer model, the thermal balance strategy was adopted in this study. As observed earlier, during validation of the model of valves, it may be suggested that irrespective of the valve or piston, in both cases the heat exchange equilibrium is maintained by the constant heat

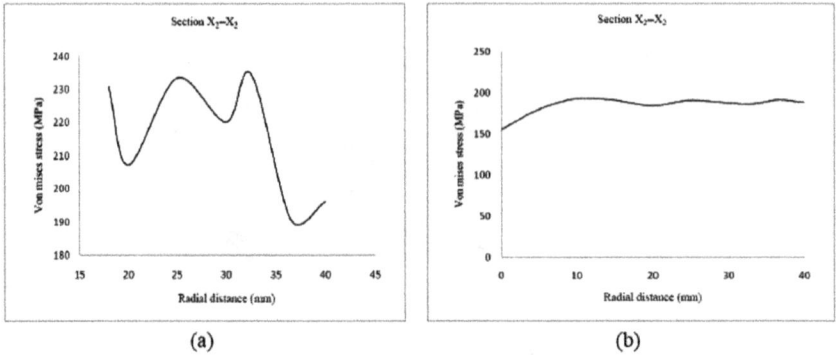

Figure 7.10 Von Mises stress distribution in conventional piston body at (a) section X₁–X₁ and (b) section X₂–X₂.

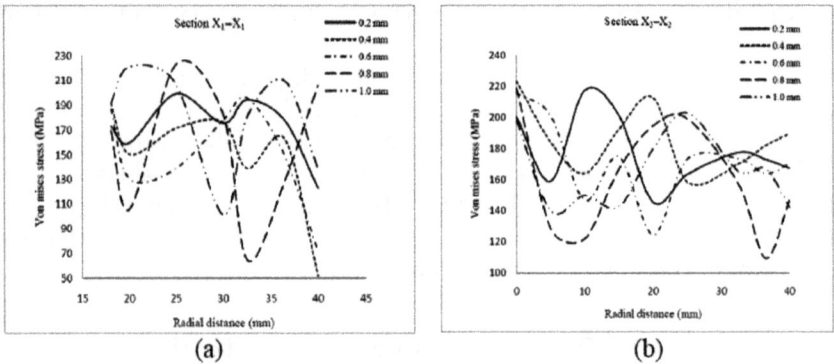

Figure 7.11 Von Mises stress in TBC piston body with 0.2, 0.4, 0.6, 0.8, and 1.0 mm coating thickness at (a) section X₁–X₁ and (b) section X₂–X₂.

entry from the head side and its release to the surroundings through air/ water. Therefore, it may be considered that the heat balance equation complies with both conventional and TB coated pistons. The degree of heat transfer across the piston with and without thermal coating is indicated in the Figure 7.12. This is evident in conventional pistons [29] that 16% of supplied heat is rejected through air and 84% through a water-cooling system. The same effect was observed in thermal barrier coated pistons with different coating thicknesses. It is also observed that after the combustion of fuel, the heat supplied to the conventional piston through combustion gases is a maximum of 2023.9 Kcal/hr and it decreases continuously from 3.4 to 12.4% as the coating thickness increases from 0.2 to 1.0 mm. The piston

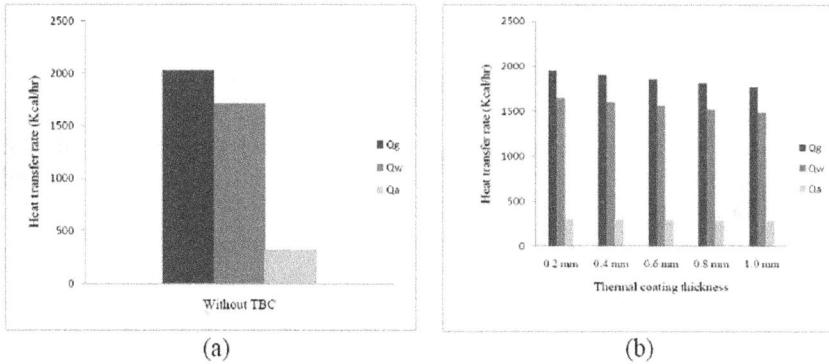

Figure 7.12 Heat transfer rate in piston (a) without coating and (b) with 0.2, 0.4, 0.6, 0.8, and 1.0 mm coating thickness [29].

protection using thermally insulating layers not only enhances the overall output of the piston but also provides extra advantages in terms of energy savings.

7.5 CONCLUSIONS

The analysis of the heat balance in the piston shows that it gives almost perfect heat balance, indicating that heat received by the hot gases is equal to heat lost to the cooling medium. Thus, the correctness of the temperature field obtained is assured. The stresses in the piston have been obtained at two different cross sections (X_1–X_1 and X_2–X_2). The analysis shows that in all the cases, the radial, tangential, axial, and von mises stresses obtained with thermal barrier coatings are lower than the corresponding values for uncoated components. The variation of radial and axial displacements has also been shown. On the basis of the results presented, the following conclusions have been drawn:

1. The maximum temperature values for the case of pistons (260.29 °C at X_1–X_1 and 265.65 °C at X_2–X_2) were found to be much lower than those pertaining to valves. In the case of engine pistons, TBC was found to cause maximum temperature drops of 23 °C and 25 °C at X_1–X_1 and X_2–X_2, respectively. The effectiveness of TBC was found to increase with increasing coating thickness.
2. For conventional pistons, the maximum radial displacement is found to be 0.175 mm and it decreases with respect to coating thickness. This reduction in radial displacement attributed to TBC ranges from 2.56 to 8.89% at X_1–X_1 and from 2.45 to 8.96% at X_2–X_2 for the range of coating thicknesses considered here.

3. Thermal barrier coatings are found to be highly effective in reducing thermal stress. This is evident from a decrease in the maximum value of von-mises stress from 234.55 to 220.82 MPa. Further, it is found that 0.6 mm thick TBC yields minimal variation in the value of von-mises stress.

4. After the combustion of fuel, the heat supplied to the conventional piston through combustion gases is found to be maximum, and it decreases continuously from 3.4 to 12.4% as the TBC thickness increases from 0.2 to 1.0 mm. Therefore, besides reduction of high temperature and thermal stresses, the use of thermal barrier coatings also leads to considerable energy saving.

REFERENCES

[1] Koff, B. L. (2003). Gas Turbine Technology Overview-A Designer's Perspective. In AIAA/ICAS International Air and Space Symposium and Exposition: The Next 100 Years (pp. 2003–2722). Dayton, OH, USA: AIAA.

[2] Eichelberg, G. (1939). Some new investigations on old combustion engine problems. Engineering, 148, 547-550.

[3] Shayler, P. J., Colechin, M. J. F., & Scarisbrick, A. (1996). Heat transfer measurements in the intake port of a spark ignition engine. SAE Transactions, 257-267.

[4] Cooper, M. G., Mikic, B. B., & Yovanovich, M. M. (1969). Thermal contact conductance. International Journal of Heat and Mass Transfer, 12(3), 279-300.

[5] Engh, G. T., & Chiang, C. (1970). Correlation of convective heat transfer for steady intake-flow through a poppet valve (No. 700501). SAE Technical Paper.

[6] Goldstein, R. J., Eckert, E. R. G., Ibele, W. E., Patankar, S. V., Simon, T. W., Kuehn, T. H., and Srinivasan, V. (2005). Heat transfer-a review of 2002 literature. International Journal of Heat and Mass Transfer, 48(5), 819-927.

[7] Abu-Nada, E., Al-Hinti, I., Al-Sarkhi, A., & Akash, B. (2006). Thermodynamic modeling of spark-ignition engine: Effect of temperature dependent specific heats. International Communications in Heat and Mass Transfer, 33(10), 1264-1272.

[8] Assanis, D. N., & Mathur, T. (1990). The effect of thin ceramic coatings on spark-ignition engine performance. SAE Transactions, 981-990.

[9] Heywood, J. B. (1988). Internal combustion engine fundamentals. McGraw-Hill Book Company. St. Louis, MO, 230, 245.

[10] Furuhama, S., & Enomoto, Y. (1987). Heat transfer into ceramic combustion wall of internal combustion engines. SAE Transactions, 38-53.

[11] Kamo, R., Assanis, D. N., & Bryzik, W. (1989). Thin thermal barrier coatings for engines. SAE Transactions, 131-136.

[12] Jaichandar, S., & Tamilporai, P. (2003). Low heat rejection engines-an overview.

[13] Cerit, M., & Coban, M. (2014). Temperature and thermal stress analyses of a ceramic-coated aluminum alloy piston used in a diesel engine. International Journal of Thermal Sciences, 77, 11-18.

[14] Woschni, G., Spindler, W., & Kolesa, K. (1987). Heat insulation of combustion chamber walls-a measure to decrease the fuel consumption of IC engines? SAE Transactions, 269-279.

[15] Woschni, G. (1967, September). Paper 8: Electronic Calculation of the Time Curve of Pressure, Temperature and Mass Flow Rate in the Cylinder of a Diesel Engine. In Proceedings of the Institution of Mechanical Engineers, Conference Proceedings (Vol. 182, No. 12, pp. 71-78). Sage UK: London, England: SAGE Publications.

[16] Woschni, G. (1967). A universally applicable equation for the instantaneous heat transfer coefficient in the internal combustion engine (No. 670931). SAE Technical paper.

[17] Shalev, M., Zvirin, Y., & Stotter, A. (1983). Experimental and analytical investigation of the heat transfer and thermal stresses in a cylinder head of a diesel engine. International Journal of Mechanical Sciences, 25(7), 471-483.

[18] Wu, H. W., & Chiu, C. P. (1989). Finite element model for thermal system in real time operation diesel piston. Computers & Structures, 32(5), 997-1004.

[19] Ong, J. H. (1990). Steady state thermal analysis of a diesel engine piston. Computers in Industry, 15(3), 255-258.

[20] Prasad, R., & Samria, N. K. (1990). Transient heat transfer analysis in an internal combustion engine piston. Computers & Structures, 34(5), 787-793.

[21] Parlak, A., & Ayhan, V. (2007). Effect of using a piston with a thermal barrier layer in a spark ignition engine. Journal of the Energy Institute, 80(4), 223-228.

[22] Nakic, D. J., Assanis, D. N., & White, R. A. (1994). Effect of elevated piston temperature on combustion chamber deposit growth. SAE Transactions, 1454-1466.

[23] Poola, R. B., Nagalingam, B., & Gopalakrishnan, K. V. (1994). Performance of thin-ceramic-coated combustion chamber with gasoline and methanol as fuels in a two-stroke SI engine (No. ANL/ES/CP-83720; CONF-9410173-2). Argonne National Lab., IL (United States).

[24] Moughal, K. S., & Samuel, S. (2007). Exhaust emission level reduction in two-stroke engine using in-cylinder combustion control. SAE Transactions, 668-672.

[25] Buyukkaya, E., & Cerit, M. (2008). Experimental study of NOx emissions and injection timing of a low heat rejection diesel engine. International Journal of Thermal Sciences, 47(8), 1096-1106.

[26] Banka, V. K., & Ramesh, M. R. (2018). Thermal analysis of a plasma sprayed ceramic coated diesel engine piston. Transactions of the Indian Institute of Metals, 71(2), 319-326.

[27] Binder, C., Nada, F. A., Richter, M., Cronhjort, A., & Norling, D. (2017). Heat loss analysis of a steel piston and a YSZ coated piston in a heavy-duty diesel engine using phosphor thermometry measurements. SAE International Journal of Engines, 10(4), 1954-1968.

[28] Sharma, S. K., Saini, P. K., & Samria, N. K. (2015). Computational Modeling of Temperature Field and Heat Transfer Analysis for the Piston of Diesel Engine with and without Air Cavity. Jordan Journal of Mechanical & Industrial Engineering, 9(2).

[29] Sharma, S. K., Saini, P. K., & Samria, N. K. (2015). Experimental thermal analysis of diesel engine piston and cylinder wall. Journal of Engineering, 2015, http://dx.doi.org/10.1155/2015/178652.

Chapter 8

Effect of parametric variation on surface roughness of EN-8 steel in grinding

Dinesh Kumar Patel, Ajay Kumar Jena, and Sachin Kumar

8.1 INTRODUCTION

The grinder is a multipoint cutting tool that uses a hard abrasive medium to machine the surfaces. In the grinding process, thousands of abrasive cutting points of a grinding wheel are engaged simultaneously during the machining of a workpiece. Due to the cutthroat competition in the manufacturing industry, particularly in machining, the advanced technique is required in order to achieve a good-quality finished product. Sangwan et al. (2015) revealed that the heat generated during the grinding process can be eliminated by using the proper quantity of coolant at the right flow rate. The change in the grinding process parameters has been observed to produce a large amount of heat. Hence surface roughness is improved by controlling the heat generated during the operation. Sim et al. (2006) predicted that the surface roughness of steel can be improved by using the RSM model. This model is used to predict surface roughness by considering the output response obtained from the experiment. To achieve the geometrical accuracy and quality of components is a stupendous task. Jagtap et al. (2011) proposed that the optimum result obtained from the grinding

DOI: 10.1201/9781003360001-8

operation can be used by manufacturers to select a suitable combination of grinding input parameters to achieve a smooth surface with reasonable dimensional tolerance. The input parameters selected for experiments are work speed, speed of wheel, depth of cut, and number of passes. Mahata et al. (2020) analyzed the experimental findings obtained from a grinding machine by the empirical mode decomposition (EMD) method and predicted the roughness value. The time domain features of the raw signals predicted 48% accuracy. The use of marginal spectra in time domain improves the accuracy greater than 88%. Hassui et al. (2003) investigated the roughness value on the basis of tool wear. The vibration in the grinding wheel is due to the grinding wheel wear. If wear is greater, vibration is greater, and quality of product is poor. The arrangement was such that the quality of a product can be obtained even if the vibration is much higher than that obtained with a dressed wheel. The effect of vibration produced during grinding has no significant impact on surface finish. Winter et al. (2014) performed an experiment and observed optimal results in the internal cylindrical grinding process. Empirical models were created to analyze the grinding processes. Multiobjective optimizations were put to use, where weighted max-min and a geometric programming model were used respectively. Kishna et al. (2006) applied a scatters search approach of optimization to investigate the grinding input process parameters. The input parameters taken are lead of dressing, depth of dressing, workpiece speed, and speed of wheel. A mathematical model was also developed to analyze the experimental data. The results were compared with Ant Colony Optimization, quadratic programming techniques, and Genetic Algorithm. Midha et al. (1991)put forward a computerized system for the analysis of the process parameters in order to get the best practical combination of input data. A common grinding model structure was also developed by using the data from the experimental results. The optimal process conditions as determined by analysis were matched with experimental results. Wang et al. (2002) developed a mathematical model for finding the relation among the response and input parameters. With the use of computer simulation, nonlinear input parameters were generated. Real grinding input parameters were controlled by using the results of the simulation. The results derived by conducting experiments were analyzed and compared with optimum models. Results obtained satisfied the established optimum model. Kumar et al. (2015) concluded that workpiece speed was the dominating factor in the analysis of the roughness. The contribution of workpiece speed, among all other grinding input parameters, was close to 40%. The percentage error between the predicted and experimental values was 2.94%, which was very low. Singh et al. analyzed the input parameters of cylindrical grinding. The workpiece material on which experiment was conducted was AISI 4140 steel. The dominating factor in surface roughness analysis was depth of cut. The percentage contribution of depth of cut was 29.21%; second input parameter that dominated after depth of cut was workpiece speed at 26.50%. Wheel speed contributed least among all input parameters. Lijohn et al. (2013) compared the regression method and the Taguchi method for surface roughness. The roughness value by the Taguchi method and regression method were 0.47 and 0.469Ra, respectively. The optimum input data obtained from analysis of experiments are depth of cut, hardness, and speed. The confirmation test on the basis of these optimum input data gave the minimum surface roughness. Unsacar et al. (2005) conducted an experiment on a CNC grinding machine, taking AISI1050 steel as a workpiece material. The input important parameters such as work speeds, feeds, materials of wheel, wheel dressing, vibrations of the floor, and coolant used all influence the quality of the workpiece. The results of the experiments were analyzed by analysis of variance for percentage contribution of the input parameters used. The most influencing parameters in surface roughness were work speed at 47.69 and depth of cut at 39.9%. Chenet al. (2015) made a relationship between SR AND SSD depths. They also dictated the influence of grinding input parameters on SR and SSD depth. The percentage contribution of wheel speed was greater than that of all other input parameters involved in SR and SSD depth. The second most affecting parameter was feed. The increase of wheel speed and decrease of feed lower the SR and SSD depth. Many researches optimize the grinding parameters during grinding to minimize surface roughness. The modeling the real-time design of experiments was adopted, and the tests were performed as per the design. The ANOVA approach is used in order to find out the contribution of different input parameters involved in the experiments.

8.2 CYLINDRICAL GRINDING

The cylindrical grinding is a finishing operation of the workpiece. In cylindrical grinding, the workpiece rotates about the polar axis, whereas the surfaces ground are concentric to the polar axis. Cylindrical grinding is the only economical machining method for hard steel and suitable for light work. Grinding depth is determined by the feed of the wheel perpendicular to polar axis. The desired input parameters in the workpiece include depth of cut, speed of work, material of grinding wheel used, and speed of the grinding wheel. Applications of this grinding method include finishing bearing diameters, hard hydraulic cylinder shafts, hard-faced retainer pins, transmission shafts, and sizing of hardened axles. However, various research has been done on cylindrical grinding to optimize the quality of product in terms of surface roughness in accordance with the input variables. In this chapter, in-depth analysis of various parameters is done so as to achieve a quality surface finish. The optimization technique is used to find out the best possible combination of input parameters for achieving a desired roughness value.

8.3 EXPERIMENTATION

The workpiece EN8 steel is taken as the workpiece for conducting these experiments due to its versatile application in industries like shafts, studs, spindles, connecting rods, and screws, as it has excellent mechanical properties. The tensile strength varies in the range of 500–800N/mm². The cylindrical rod of EN8 steel was first center-drilled and turned on a lathe machine for rough machining. The final machining of cylindrical bars was carried out on a cylindrical grinding machine as per the L16 orthogonal array. The surface roughness values were taken by using MITUTOYO surf test SJ-400.

8.4 MODELING OF PROCESS VARIABLES AND THEIR LEVELS

To conduct the experiment, a round bar of 100 mm long and 15 mm diameter has been taken. First the workpiece is turned on the central lathe machine. The workpiece is turned to get the same size for all rods, so that the circumferential speed should not vary at all. Aluminum oxide is taken as grinding wheel material. For experimentation, five input parameters were taken: depth of cut (A), table feed (B), work speed (C), grinding wheel speed (D), and coolant (E).Coolant and grinding wheel speed were taken at two levels, and all other parameters were taken at four levels. Table 8.1 shows grinding input parameters and their corresponding levels.

8.4.1 Taguchi methods

This is a statistical method of optimization that uses orthogonal array for conducting experiments. The orthogonal array provides minimum number of test that is to be performed with high level of accuracy and to provide a set of balanced experimental tables and desired output. This ensures the backbone for prediction of optimization and data analysis. The experimental results are utilized to produce the signal-to-noise ratios.

Table 8.1 Factors and levels

Factors	Process parameters	Level1	Level2	Level3	Level4
A	Depth of cut (mm)	0.015	0.025	0.035	0.045
B	Table feed (m/min)	0.01	0.015	0.02	0.025
C	Work speed (rpm)	130	280	325	560
D	Grinding wheel speed (rpm)	1400	2800	–	–

Table 8.2 L16 orthogonal array

S. no.	Depth of cut (mm)	Table feed (m/min)	Work speed (rpm)	Grinding wheel speed (rpm)	Coolant
1	1	1	1	1	1
2	1	2	2	1	1
3	1	3	3	2	2
4	1	4	4	2	2
5	2	1	2	2	2
6	2	2	1	2	2
7	2	3	4	1	1
8	2	4	3	1	1
9	3	1	3	1	2
10	3	2	4	1	2
11	3	3	1	2	1
12	3	4	2	2	1
13	4	1	4	2	1
14	4	2	3	2	1
15	4	3	2	1	2
16	4	4	1	1	2

In the present chapter, L16 orthogonal array is considered as shown in Table 8.2.

8.5 RESULTS

The experiments were conducted on a grinding machine, and its process variables were analyzed with Taguchi L16 orthogonal array, as given in Table 8.2. The surface roughness values measured by using the MITUTOYO surf test SJ-400 and the obtained values of roughness (Ra) of the grinding process parameters are given in Table 8.3. These values are further used to find the S/N ratio.

8.5.1 Evaluation of S/N ratios

The Taguchi method uses a loss function to determine the quality characteristics. The value of output response surface roughness is converted to a signal-to-noise ratio. The Taguchi method uses three types of performance characteristic. The first one is the nominal-the-better, the second is the higher-the-better, and the third is lower-the-better. The optimum level of the input parameters is obtained by taking the highest signal-to-noise ratio. The lower-the-better criterion for surface roughness was selected for obtaining optimum grinding performance characteristics. Based on these values, the graph between the input parameter and the S/N ratio is plotted. This graph

Table 8.3 Surface roughness values as per L16 orthogonal array

S. no.	Depth of cut (mm)	Table feed (m/min)	Work speed (rpm)	Grinding wheel speed (rpm)	Coolant	Surface roughness(Ra)
1	0.015	0.01	130	1400	Dry	0.9265
2	0.015	0.015	280	1400	Dry	0.8373
3	0.015	0.02	325	2800	Flooded	0.7843
4	0.015	0.025	560	2800	Flooded	1.1025
5	0.025	0.01	280	2800	Flooded	0.7954
6	0.025	0.015	130	2800	Flooded	0.7682
7	0.025	0.015	560	1400	Dry	1.0957
8	0.025	0.025	325	1400	Dry	0.9436
9	0.035	0.01	325	1400	Flooded	0.7652
10	0.035	0.015	560	1400	Flooded	0.9835
11	0.035	0.02	130	2800	Dry	1.1282
12	0.035	0.025	280	2800	Dry	1.3475
13	0.045	0.01	560	2800	Dry	1.5645
14	0.045	0.015	325	2800	Dry	1.2463
15	0.045	0.02	280	1400	Flooded	0.7965
16	0.045	0.025	130	1400	Flooded	1.5684

depicts the variation trend of the output response with respect to the input. The S/N ratio is calculated by using the following equation:

$$S/N = -10\log 10\left(\frac{\sum y^2}{n}\right)$$

The obtained values of S/N ratios are given in Table 8.4.

8.5.2 Level mean response analysis

The level mean values help in analyzing the trend of the output response with respect to the variation of input parameters. The optimum values of input parameters are selected by taking the maximum values of the signal-to-noise ratio into consideration. The delta is calculated by taking the difference between the greatest and the lowest values of the S/N ratios. The greater value of delta among the input parameters is placed in the first rank. In a similar manner, other input parameters are ranked accordingly, as shown in Table 8.5. The responses obtained were plotted for surface roughness based on the S/N ratios in the grinding operation. A greater value of the S/N ratio is desirable for input parameters. The depth of cut at level 2 was greater, which is considered for better surface finish. The S/N ratio

Table 8.4 S/N values for grinding the EN8 steel

S. no.	Depth of cut (mm)	Table feed (m/min)	Work speed (rpm)	Grinding wheel speed (rpm)	Coolant	Surface roughness(Ra)	S/N ratio
1	0.015	0.010	130	1400	Dry	0.9265	0.0000
2	0.015	0.015	280	1400	Dry	0.8373	-1.5836
3	0.015	0.020	325	2800	Flooded	0.7843	-0.1720
4	0.015	0.025	560	2800	Flooded	1.1025	-3.5218
5	0.025	0.010	280	2800	Flooded	0.7954	-4.0824
6	0.025	0.015	130	2800	Flooded	0.7682	-5.1054
7	0.025	0.015	560	1400	Dry	1.0957	0.0000
8	0.025	0.025	325	1400	Dry	0.9436	-3.5218
9	0.035	0.010	325	1400	Flooded	0.7652	-1.9382
10	0.035	0.015	560	1400	Flooded	0.9835	-0.1720
11	0.035	0.020	130	2800	Dry	1.1282	-1.5836
12	0.035	0.025	280	2800	Dry	1.3475	0.0000
13	0.045	0.010	560	2800	Dry	1.5645	10.4575
14	0.045	0.015	325	2800	Dry	1.2463	6.0206
15	0.045	0.020	280	1400	Flooded	0.7965	13.9794
16	0.045	0.025	130	1400	Flooded	1.5684	10.4576

Table 8.5 ANOVA for S/N ratios in grinding processes by taking 95% confidence level

Level	Depth of cut (mm)	Table feed (m/min)	Work speed (rpm)	Grinding wheel speed (rpm)	Coolant
1	0.8671	0.2721	-0.5008	0.3065	-0.9403
2	0.9973	0.5162	0.7291	-0.4871	0.7597
3	-0.2923	0.5613	0.7567	–	–
4	-1.9332	-1.7108	-1.3461	–	–
Delta	2.9305	2.2720	2.1028	0.7936	1.7000
Rank	1	2	3	5	4

of table feed level3, work speed level3, grinding wheel speed level1, and coolant level1 were greater, and that gives a better surface finish.

8.5.3 Analysis of variance

Analysis of variance is a tool to find the most influencing input parameter to the surface roughness. Table 8.6 shows the result of ANOVA for the grinding operation. The table is drawn by considering a 95% confidence level. The dominating factor in the experiment was depth of cut with a total 36.27% of contribution. The contribution of other input parameters on

Table 8.6 Result of ANOVA for grinding operation

Source	DF	Seq SS	Adj SS	Adj MS	F	P	SS %
Depth of cut (mm)	3	0.40140	0.40140	0.13380	4.62	0.087	36.27
Table feed (m/min)	3	0.22164	0.22164	0.07388	2.55	0.194	20.02
Work speed (rpm)	3	0.18023	0.18023	0.06008	2.08	0.246	16.28
Grinding wheel speed (rpm	1	0.04205	0.04205	0.04205	1.45	0.295	3.80
Coolant	1	0.14547	0.14547	0.14547	5.03	0.088	13.15
Residual error	4	0.11577	0.11577	0.02894			
Total	15	1.10655					

Table 8.7 Optimum control parameters for minimum surface roughness

Depth of cut (mm)	Table feed (m/min)	Work speed (rpm)	Grinding wheel speed (rpm	Coolant	S/N ratio
A0	B0	C0	D0	E0	–
0.025	0.02	325	1400	Flooded	–1.4269

surface roughness are table feed (20.02%), work speed (16.28%), grinding wheel speed (3.8%), and coolant (13.15%) in grinding the workpiece.

8.5.4 Mathematical modeling

The output response, i.e. surface roughness corresponding to the optimum level, was calculated by taking the optimum combination into consideration. In Table 8.7, the optimum level of the input parameter is given. The value of S/N ratios was calculated by equation (8.1).

$$H= ή+ (A0 –ή) + (B0 –ή) + (C0 –ή) + (D0 –ή) + (E0 –ή) \quad\quad (8.1)$$

The symbol η is the S/N ratio at the optimum level, and ή is the mean S/N ratios of all parameters. A0 (0.9973), B0 (0.5613), C0 (0.7567), D0 (0.3065), and E0 (0.7597) are the mean S/N ratios. All the values are placed in equation (8.1), and the S/N ratio is calculated. This predicted value was –1.4269 dB, which is further transformed in average roughness Ra (0.7737 μm).

8.5.5 Confirmation test

To validate the result, a confirmation test was conducted by taking the optimum input parameters. The confirmation experiments were performed at the optimum variable level (i.e. depth of cut 0.025 mm, table feed 0.02 m/min, work speed 325 rpm, grinding wheel speed1400 rpm, and coolant flooded). The value of surface roughness in experiment was found to be 0.7845 µm. The result is very close to the predicted value of surface roughness as achieved by the Taguchi method. It was concluded that there was an error of 1.39% between the predicted value and experimental value.

8.6 Conclusions

The chapter proposes an integrated optimization approach. The optimum level was calculated by taking the S/N ratio in calculation. These values are depth of cut 0.025 mm, table feed 0.02 m/min, work speed 325 rpm, grinding wheel speed1400 rpm, and coolant flooded. It has been found that the depth of cut contributes more than 36% in the minimization of surface roughness, whereas table feed and work speed affect the surface roughness to 20.08 and 16.28%, respectively. The predicted value of surface finish is 0.7737 µm at the 95% confidence level. The error percentage in the final calculated value is due to the presence of vibration in the machine tool and noise factors. The percentage error of surface roughness was found to be 1.39% for the cylindrical grinding process. The most dominating factor in machining was depth of cut, followed by table feed.

REFERENCES

1. Patel, D.K., Goyal, D., Pabla, B.S. (2018). Optimization of parameters in cylindrical and surface grinding for improved surface finish, Royal Society Open Science, 5, 1–11.
2. Sangwan, K.S., Saxena, S., Kant, G. (2015). Optimization of machining parameters to minimize surface roughness using integrated ANN-GA approach, Procedia CIRP, 29, 305–310.
3. Lil, G.F., Wang, S., Yang, L.B. (2002). Multiparameter optimization and control of the cylindrical grinding process, Journal of Materials Processing, Technology, 129, 232–236.
4. Tawakoli, T., Rabiey, M. (2007). Dry grinding by special conditioning, Int. J. Adv. Manuf. Technology, 33, 419–424.
5. Sim, S.B., Kwak, J.S., Jeong, Y.D. (2006). Analysis of grinding power and surface roughness in external cylindrical grinding of hardened SCM 440 steel using the response surface method, International Journal of Machine Tools & Manufacture, 46, 304–312.
6. Jagtap, K.R., Ubale, S.B., Kadam, M.S. (2011). Optimization of cylindrical grinding process parameters for AISI 5120 Steel using Taguchi Method, International Journal of Design and Manufacturing Technology, 2, 47–56.
7. Mahata, S., Shaka, P., Babu, N.R., Pradeep, K., Prakasam, P.K. (2020). In process characterization of surface finish in cylindrical grinding process using vibration and power signals, Procedia CIRP, 88, 335–340.

8. Hassui, A., Diniz, A.E. (2003). Correlating surface roughness and vibration on plunge cylindrical grinding of steel, International Journal of Machine Tools and Manufacture, 8, 855–862.

9. Winter, M., Lie, W., Kara, S., Herrmann, C. (2014). Determining optimal process parameters to increase the eco-efficiency of grinding processes, Journal of Cleaner Production, 66, 644–654.

10. Li, G.F., Wang, L.S. Yang, L.B. (2002). Multi-parameter optimization and control of the cylindrical grinding process, Journal of Material Process Technology, 129, 232–236.

11. Krishna, A.G., Rao, K.M. (2006). Multi-objective optimisation of surface grinding operations using scatter search approach, Int. J. Adv. Manuf. Technology, 29, 475–480.

12. Midha, P.S., Zhu, C.B., Trmal, G.J. (1991). Optimum selection of grinding parameters using process modeling and knowledge based system approach, Journal of Material Process Technology, 28, 189–198.

13. Lee, C.W., Choi, T., Shin, Y.C. (2003). Intelligent model-based optimization of the surface grinding process for heat-treated 4140 steel alloys with aluminum oxide grinding wheels, J. Manuf. Sci. Eng., 125, 65–76.

14. Guo, C., Malkin, S. (2001). Cylindrical Grinding Process Simulation, Optimization, and Control, Society of Manufacturing Engineers, ME MR01-334.

15. Wang, L.S., Li, G.F., Yang, L.B. (2002). Multi-parameter optimization and control of the cylindrical grinding process, Journal of Materials Processing Technology, 129, 232–236.

16. Kumar, S., Subramaniyan, S.D. (2018). A review of cylindrical grinding process parameters by using various optimization techniques and their effects on the surface integrity, wear rate and MRR, International Journal of Advance Engineering and Research Development, 5, 719–729.

17. Jagtap, K. R., Ubale, S.B., Kadam, M.S. (2011). Optimization of cylindrical grinding process parameters for AISI 5120 steel using Taguchi method, International Journal of Design and Manufacturing Technology, 2, 47–56.

18. Kumar, K. (2012). Optimal material removal and effect of process parameters of cylindrical grinding machine by Taguchi method, International Journal of Advanced Engineering Research and Studies, 2, 39–43.

19. Kumar, S., Bhatia, O.S. (2015). Review of analysis and optimization of cylindrical grinding process parameters on material removal rate of EN15AM steel, Journal of Civil and Mechanical Engineering, 1, 35–43.

20. Manickam, M.M., Kalaiyarasan, V. (2014). Optimization of cylindrical grinding process parameters of OHNS steel (AISI O-1) rounds using Design of Experiments concept, International Journal of Engineering Trends and Technology, 17, 109–114.

21. Kumar, N., Tripathi, H. (2015). Optimization of cylindrical grinding process parameters on C40E steel sing Taguchi Technique, International Journal of Engineering. Research & Applications, 5, 100–104.

22. Singh, K., Kumar, P., Goyal, K. (2014). To study the effect of input parameters on surface roughness of cylindrical grinding of heat treated AISI 4140 steel, American Journal of Mechanical Engineering, 2, 58–64.

23. Lijohn, G.P., Job, K.V., Chandran, L.M. (2013). Study on surface roughness and its prediction in cylindrical grinding process based on Taguchi method of optimization, International Journal of Engineering Development and Research, 2, 2486–2491.

24. Unsacar, F., Saglam, Yaldiz, S. (2005). An experimental investigation as to the effect of cutting parameters on roundness error and surface roughness in cylindrical grinding, International Journal of Production Research, 43, 2309–2322.

25. Chen, J., Fang, Q., Li, P. (2015). Effect of grinding wheel spindle vibration on surface roughness and subsurface damage in brittle material grinding, International Journal of Machine Tools & Manufacture, 91, 12–23.

Chapter 9

Effect of REOs on tribological behavior of aluminum hybrid composites using ANN

Vipin Kumar Sharma, Vinod Kumar,
Ravinder Singh Joshi, and Ashwani Kumar

9.1 INTRODUCTION

The demand of aluminum alloys increased day by day in the era of Industry 4.0 due to their excellent mechanical properties. These alloys are widely used in every sector of engineering due to their better load bearing capacity, capability to reduce oxide formation on a targeted surface, good thermal resistance, and good strength-to-weight ratio [1, 2]. Because of such properties, aluminum alloys are well suited for the very famous applications like defense, aerospace, automotive industries, electronic industries, and medical

DOI: 10.1201/9781003360001-9

applications [3, 5]. In view of this fact, aluminum alloys have been widely used in tribological applications like aircraft housing, O-rings, liners and clutches of internal combustion engines, pivot, rockers arm, and pulleys. It is therefore very much important to focus on the wear-based study of aluminum alloys along with their techniques for improving wear properties [6, 7]. The scientific community doing ongoing research has always focused on enhancing the metallographic and mechanical characterization along with wear resistance properties of aluminum alloys. The wear properties have been improved significantly after careful examination of reinforcement in terms of its purity and size, morphology control with a suitable casting technique, weight percentage of alloying elements to be used, its capability to make a perfect bonding with the matrix, and good dispersion in matrix [8–10]. In almost all research related to the fabrication of composites via the casting process, it has been seen that the mechanical properties of the alloys have been improved with the introduction of a secondary phase called reinforcement [11, 12].

The literature has also shown that the various approaches used in the predictions of mechanical and wear behavior of hybrid AMCs have been based on experimental and data models [13]. The number of experiments has been reduced by the use of data models. Zhang et al. [14] predicted the tribological property of composites reinforced with short fibers using an artificial neural network. The predicted results are in very close agreement with the experimental results. Gyurova et al. [15] predicted the wear behavior and coefficient of friction properties of polyphenylene sulfide composites under dry sliding conditions using ANN approach. The results revealed that the predicted results of wear as well as of friction are very comparable with the experimental results.

Veeresh et al. [16] prepared the Al-7075 /SiC composites using the liquid metallurgy process and investigated the wear behavior of composites under dry sliding conditions. The weight percentage of silicon carbide varied from 2 to 6%. The result showed that the percentage increase in silicon carbide leads to an increase in wear resistance. Finally, the ANN model was developed to predict the wear behavior aluminum composite reinforced with SiC. The predicted results obtained using the artificial neural network (ANN) approach was found to be in good agreement with the measured values. Ozyurek et al. [17] analyzed the dry sliding wear behavior of aluminum alloy reinforced with silicon carbide by using both ANN and experimental results. The results of wear analysis using ANN approach have shown to offer a better predictive index as compared to the real experimental data. Genel et al. [18] explored the multiple-layer feed-forward approach for the prediction of wear behavior of aluminum composites reinforced with alumina fibers and zinc. The composites were fabricated via a pressure die casting method by varying the reinforcement from 10 to 30 wt.%. The wear rate and coefficient of friction were mostly affected by the weight percentage of reinforcement. The wear rate declines with the increment in the percentage of reinforcement and increases with the increase of normal load. Finally, the wear data of the ANN model and experimental data have been compared using three-dimensional plots. The empirical relation for wear rate and the coefficient of friction related to the applied normal load and volume fractions of reinforcement has been established.

Shirvanimoghaddam et al. [19] prepared A356.1 aluminum composites reinforced with boron carbide via the stir casting route by varying boron carbide from 5 to 15wt.%. For successfully predicting the mechanical properties of the composites, two models (Levenberg–Marquardt Algorithm neural network and thin plate splines) were constructed. The result confirmed that the thin plate splines method have shown less error as compared to the Levenberg–Marquardt Algorithm-Neural network while predicting the hardness and tensile strength of the composites. Kumar et al. [20] successfully applied the ANN model while predicting the wear behavior of CuSiC composite. The results showed that the predicted values of wear behavior by the ANN model are in the very close agreement with the experimental results. Borgaonkar and Ismail [21] successfully applied the Taguchi and ANN modeling approaches while predicting the tribological performance of MoS_2-TiO_2–coating-based composites. Results have concluded that the wear behavior gives good results via the ANN approach as compared to the Taguchi approaches. Bose and Nandi [22] proposed a multioptimization technique for analyzing the titanium-based composites via EDM parameters.

The present study aims to synthesize the aluminum hybrid composites reinforced with particulate cerium oxides via the bottom pouring stir casting system. The effect of cerium oxide on the wear properties of Al-6061 were investigated using the Levenberg–Marquardt ANN approach under varying conditions of normal load, sliding distance, and sliding velocity.

9.2 MATERIALS AND METHODS

9.2.1 Manufacturing of hybrid composites and its processing

The various samples of aluminum composites with nomenclature A, B, C, D, E, F, and G with reinforcements Al_2O_3/SiC varied from 2.5 to 7.5 wt.% and from 0.5 to 2.5 wt.% of cerium oxide (CeO_2) and were manufactured by a bottom pouring stir casting machine [23, 24]. Reinforcements with average particles size 40 μm for Al_2O_3, 220 μm for SiC, and 5μm forCeO_2 were used. The reinforcement particles' average length, width, and aspect ratio were measured using Image J, as shown in Table 9.1

Table 9.1 Morphology of various reinforcement particles

Powders	Average particle length (μm)	Average particle width (μm)	Aspect ratio	Average particle size (μm)
SiC	55.7	29.5	1.89	39
Al₂O₃	50.5	28.3	1.78	42
CeO₂	29.7	9.9	2.99	16

9.2.2 Artificial neural network: Levenberg–Marquardt Algorithm

An artificial neural network is a biological-nervous-system-inspired approach that can use mathematical tools to simulate a wide variety of complex scientific and engineering problems in the era of Industry 4.0 [25, 26]. It resembles the human brain in two aspects: (1) the knowledge is acquired by the network through a learning process, and (2) interneuron connection strengths known as synaptic weights are used to store the knowledge [26]. The machine learning model based on the ANN feature is decided in large part by means of the interconnections between artificial neurons, similarly to the ones going on in their natural counterparts in biological systems [27]. Also in composites, a positive quantity of experimental effects is required to educate a well designed neural network. After the network has discovered how to resolve the material problems, new information from the same area can then be expected without having to conduct many, lengthy experiments. The goal in using ANNs is also to conduct systematic parameter studies of the best schemes for the layout of composite substances for precise packages [27]. In the ANN model, generally two steps are essential: training and testing.

A general ANN model that has been inspired by biological models is shown in Figure 9.1(a). For wear rate analysis of composites, a machine leaning technique called the Levenberg–Marquardt Algorithm was implemented. In this approach, sigmoid transfer function is generally used to adapt individual weights and bias within the network. In the structure of the ANN model, consisting of three layers as shown in Figure 9.1(b), the input variable (sliding distance, load, and CeO_2 wt.%) to the model is shown by the input layer, output layer is used for specific wear rate, and two hidden layers are used as well.

9.2.2.1 Analysis of wear behavior of hybrid composites using ANN approach

ANN model with different layers was used while predicting the wear rate analysis of composites based on REOs (rare earth oxides). Three input

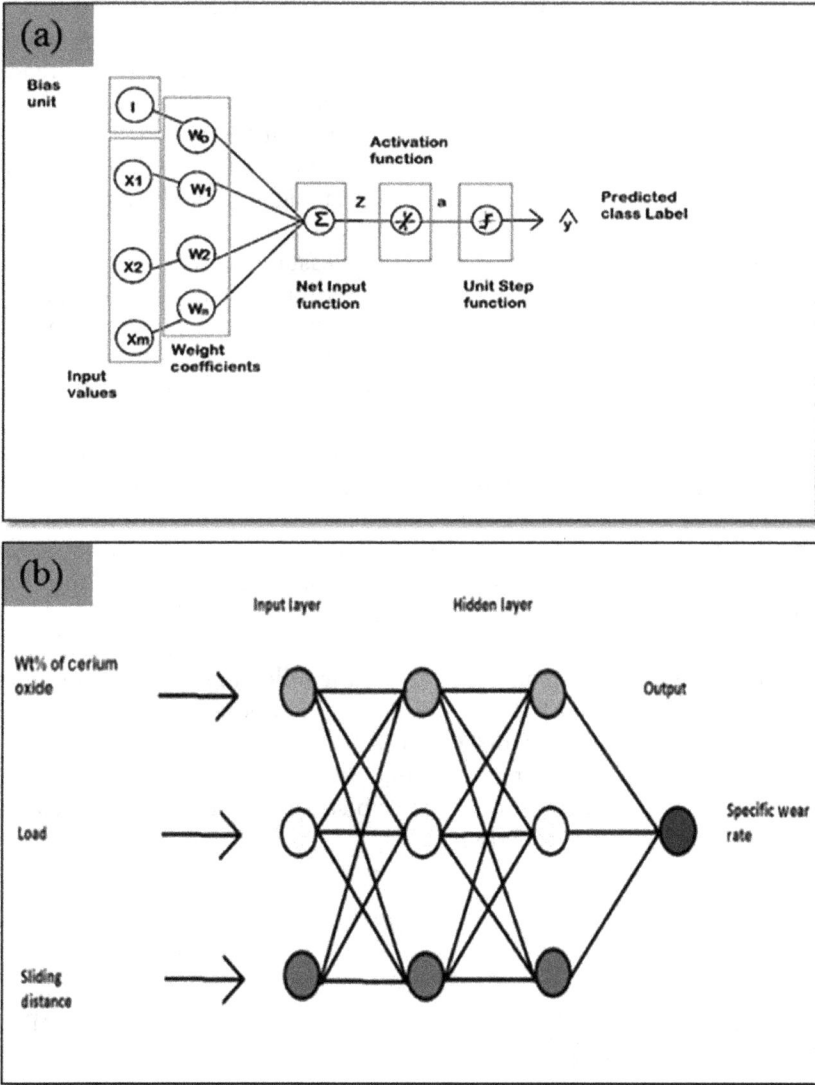

Figure 9.1 (a) Generalized ANN structure, (b) ANN network structure for REOs hybrid composites showing input, output, and hidden layers.

neuron layers with parameters –sliding distance, normal load, and weight percentage of cerium –were taken. The wear rate was described by the output layer with one neuron. A single hidden layer with 10 neurons was also used. First was the parameter sliding distance with values 1000, 1500, and 2000 m, the second parameter was normal load with values 10, 15, and 20 N, and the final parameter had varying weight percentages of cerium oxide of 0.5, 1.5, and 2.5. The hit-and-trial approach based on the mean square error criterion was implemented to determine the number of neurons in the hidden layer. A Levenberg–Marquardt back propagation (LMBP) network with single hidden layer having 10 neurons fits well in the present neural network model, and the architecture takes the form 3-[10]$_1$-1. Hidden neurons have been described by the nonlinear tangent sigmoid activation function and output neurons by linear activation function.

Wear rate of the composite is generally measured in terms of validation error in the training process. It has been determined that in the preliminary phase of training, the validation error is reduced. But when the network starts to evolve to overfit the facts, the error at the validation set generally starts to thrust upward. The validation errors increase in a distinctly wide variety of iterations. The training was stopped at the unique quantity of iterations, and the values of the weights and biases were returned to the minimum values of validation error. Finally, numerous subsets of the test were used to compare the ANN fashions. Eighty percent of the specific wear rate experimental data of REOs hybrid composites have been used for training the neural network model and 20% for validation and testing. The constructed ANN models were evaluated on the test data for the determination of the level of accuracy. These consequences of ANN, compared to the experimental information on usage, led to the following standards in order to validate the quality of wear parameter estimations: sum square error (SSE), root mean square error (RMSE), mean square error (MSE), mean relative error (MRE), maximum absolute error (MAE), regression (R), and cross-correlation coefficient (R^2). The sum square error (SSE) is calculated by using the subsequent equation:

$$\mathrm{SSE} = \sum_{i=1}^{M} = \left[W_{PI} - W_I \right]^2 \qquad (9.1)$$

The minimum value of SSE indicates better wear results. The MRE is calculated as follows:

$$\mathrm{MRE} = \sum_{i=1}^{M} = \left[W_{PI} - W_I \right]^2 / W \qquad (9.2)$$

where W_{PI} indicates the Ith predicted wear characteristics, W_I indicates the Ith measured value, and W is the mean value of W_I. The value of MRE and regression is obtained directly from the neural network training session.

9.3 RESULTS AND DISCUSSIONS

9.3.1 Wear prediction of Al-6061/SiC/Al₂O₃/REOs hybrid composites using ANN

For wear analysis of hybrid composites, the overall data set was divided into three sets: training set, validation set, and testing set. A total of 90 experiments have been carried out by varying the input parameters such as load (10, 20, and 30 N), sliding distance (1000, 1500, and 2000 m), and weight percentage of cerium oxide (0.5, 1.5, and 2.5 wt.%) in order to predict the wear behavior of Al-6061/SiC/Al₂O₃/REOs hybrid composites. Out of the overall 80 experimental data sets: 80% (64 data sets) are allocated for training, 10% (8) data sets are allocated for validation, and the remaining 10%, (8 data sets) are for testing. After that, various values of weight and bias values occur automatically during the artificial neural network function updates as per the Levenberg–Marquardt Algorithm. Then, mean square error is used for the wear prediction performance.

9.3.2 Prediction of specific wear rate for Al-6061/SiC/Al₂O₃/REOs hybrid composites

The experimental data sets shown in Table 9.2 were generally used to train the constructed 3-$[10]_1$-1 neural network model. Training, validation, and testing for various sets of data have been continued until the result reaches the minimum value of mean square error. From Table 9.2, it is identified that the multilayer network structure (3-2-1-1) gives the minimum mean square error when compared to the other number of trials at different data sets. Hence this network structure is finally used for the ANN prediction

Table 9.2 Different network trials of the ANN for prediction of specific wear rate of Al-6061/SiC/Al₂O₃/REOs hybrid composites

S.NO	Network structure	Training set MSE	Validation set MSE	Test set MSE
1	3-1-1-1	$3.10620e^{-8}$	$1.9985e^{-7}$	$1.42974e^{-8}$
2	3-2-1-1*	$3.65737e^{-10}$	$4.3424e^{-10}$	$2.49168e^{-10}$
3	3-3-1-1	$4.25500e^{-10}$	$1.08329e^{-9}$	$8.83467e^{-10}$
4	3-4-1-1	$1.35146e^{-9}$	$8.76796e^{-8}$	$2.13466e^{-8}$
5	3-5-1-1	$4.45466e^{-8}$	$3.46967e^{-8}$	$6.88280e^{-8}$
6	3-6-1-1	$3.45238e^{-8}$	$1.12878e^{-7}$	$6.51731e^{-6}$
7	3-7-1-1	$3.72411e^{-9}$	$4.25746e^{-9}$	$4.68886e^{-9}$
8	3-8-1-1	$5.11953e^{-10}$	$2.55199e^{-8}$	$7.76284e^{-8}$
9	3-9-1-1	$2.39578e^{-11}$	$8.24102e^{-7}$	$1.11998e^{-7}$
10	3-10-1-1	$2.87237e^{-10}$	$9.96801e^{-9}$	$1.27341e^{-8}$

*   The highlighted ANN network structure shows optimized values.

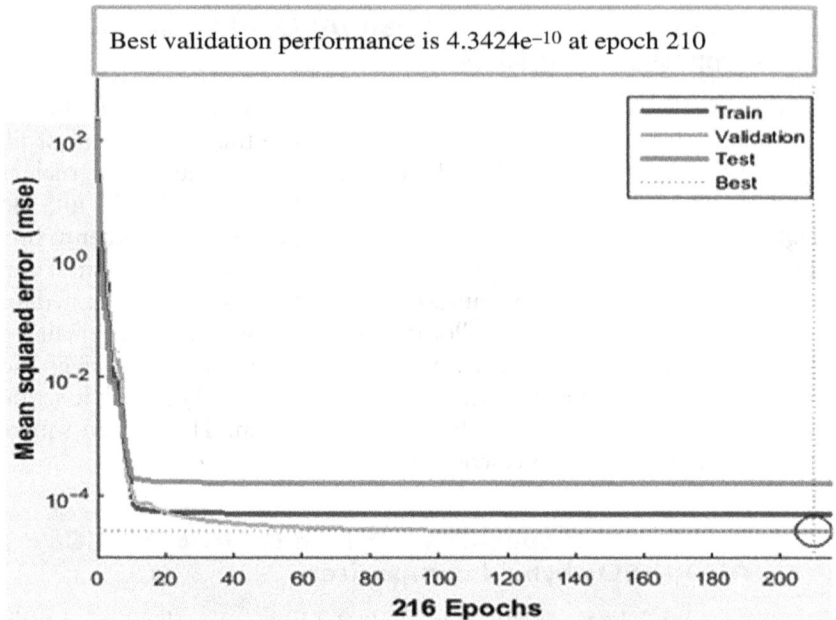

Figure 9.2 Mean squared error (MSE) of performance plot using ANN for Al-6061/SiC/Al$_2$O$_3$/REOs hybrid composites.

systems. The selected network structure of 3-2-1-1 generally indicates the relation of various input parameters with specific wear rate. The predicted results of specific wear rate using the ANN structure are very much comparable with the experimental results of wear. Figure 9.2 showed the validation MSE value as 4.3424e^{-10} for the best validation performance at 210 epochs. The errors of various trials during the experiments are shown in Figure 9.3.

9.3.3 Mean squared error (MSE) of performance plot using ANN

From Figure 9.4, it is concluded that the data show a good fit after proper training and show that the network with good training presents the output with minimum error. This output is generally used to predict the unknown value for the future. The correlation between the output value and target value is shown by the regression coefficient. When the R-value reaches 1, it means a very close relationship exists between the output and target. If R-value is zero, the random relationship exists, and if the R-value is greater than 0.9, the quality of the result is better, as stated in literatures [27].

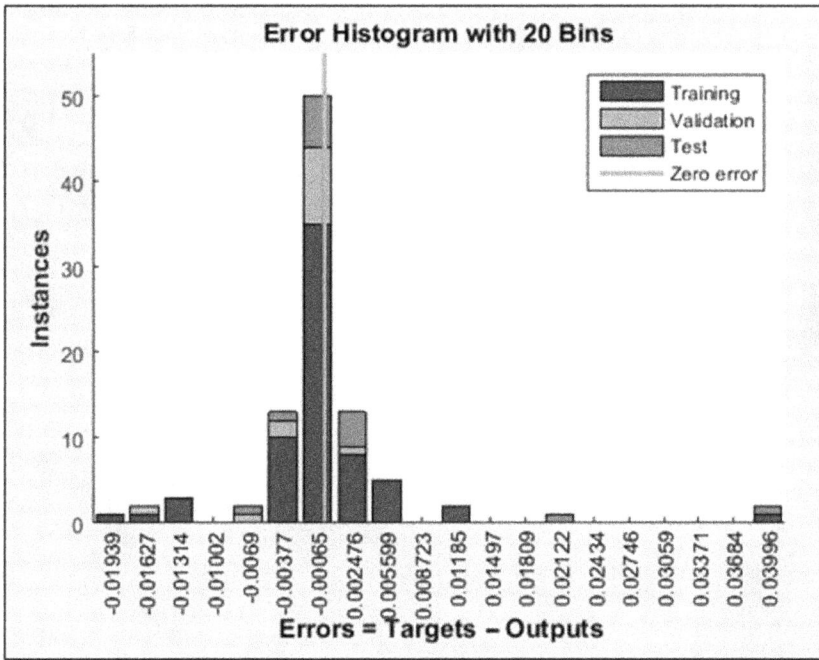

Figure 9.3 Histogram showing errors using ANN for Al-6061/SiC/Al$_2$O$_3$/REOs hybrid composites.

9.3.4 Regression plots of training, testing, and validation using ANN

The regression coefficients were obtained for training, validation, and testing using Figure 9.4. It shows a linear relation between the network output and corresponding target. The output tracks the targets very well for training, testing, and validation, and the correlation (*R*-value) is 0.98733 for the total response. This is an indication of good and smooth agreement between the experimental data and prediction of the neural network model.

9.3.5 Functional fit diagram of the training

The functional fit diagram shown in Figure 9.5 shows the training targets, training outputs, validation targets, and all the targets and outputs. All these terms fit well, as stated by the functional graph with no error.

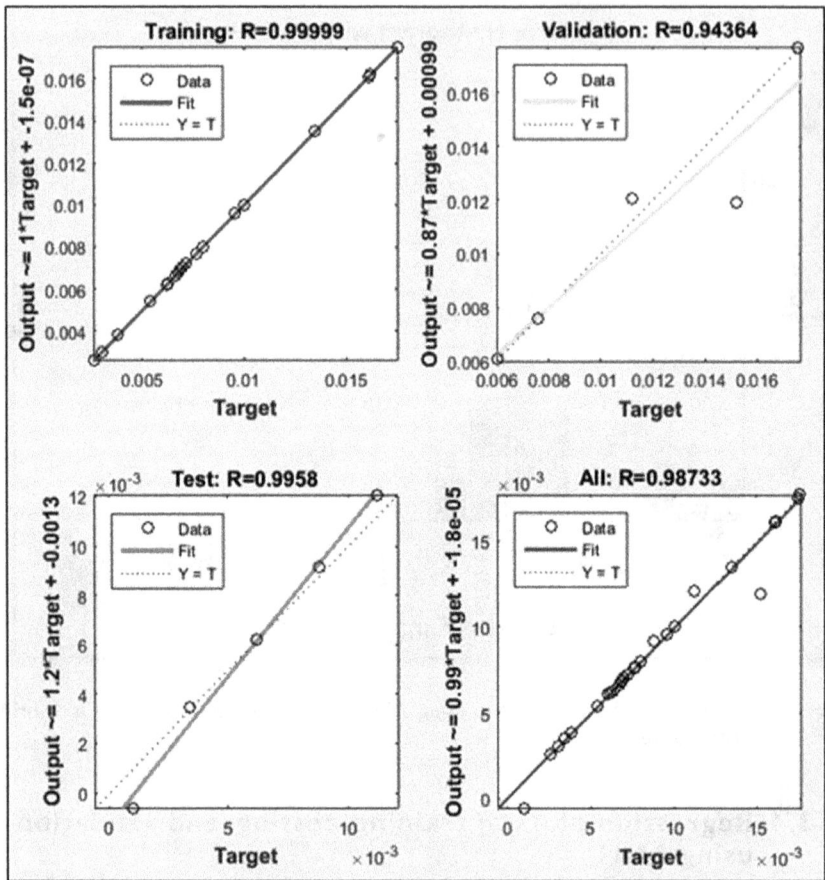

Figure 9.4 Regression plots for using ANN for Al-6061/SiC/Al$_2$O$_3$/REOs hybrid composites using Levenberg–Marquardt Algorithm.

9.3.5 Confirmation test for Al-6061/SiC/Al$_2$O$_3$/REOs hybrid composites

The confirmation test was conducted to validate the developed ANN model by preparing three specimens with different weight percentages of cerium oxide. The three prepared specimens are subjected to dry sliding wear tests by varying various input process parameters such as load, sliding distance, and wt.% of cerium oxide. The corresponding specific wear rates values are tabulated in Table 9.3. The experimental values were again tested using the already developed ANN model, as shown in Figure 9.6, for the specific wear rate. Based on the experimental and predicted results using ANN models,

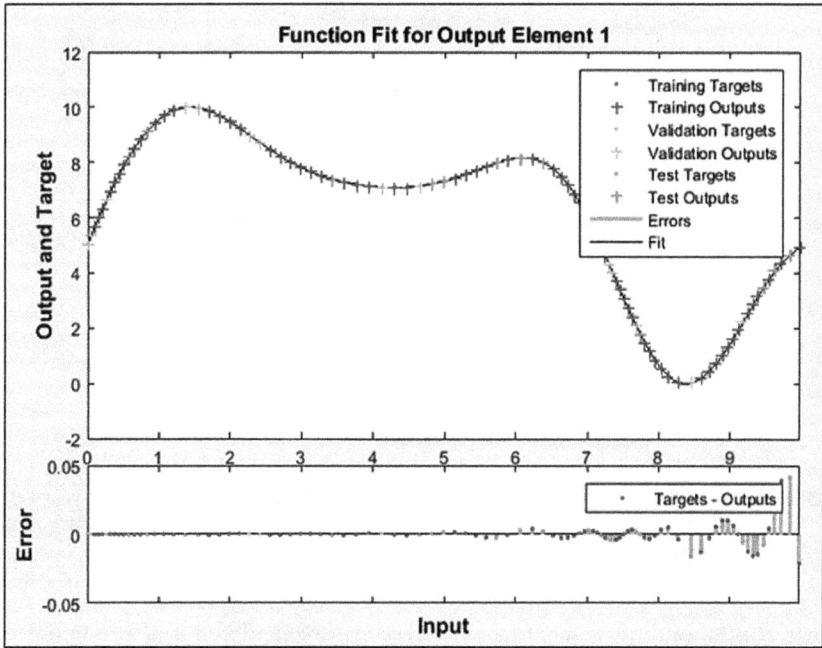

Figure 9.5 Functional fit diagram of training, testing, and validation for Al-6061/SiC/Al$_2$O$_3$/ REOs hybrid composites using Levenberg–Marquardt Algorithm.

Table 9.3 Comparison of outputs for Al-6061/SiC/Al$_2$O$_3$/REOs hybrid composites using Levenberg–Marquardt Algorithm

Ex. No.	Wt.% of CeO2	Normal load (N)	Sliding distance (m)	Specific wear rate ($\times 10^{-3}$ mm^3/nm)	
1	0.5	20	1000	Actual	0.0085
				Predicted	0.0082
				Error	3.658
2	1.5	20	1500	Actual	0.0058
				Predicted	0.0056
				Error	3.572
3	2.5	20	2000	Actual	0.0213
				Predicted	0.0209
				Error	1.914

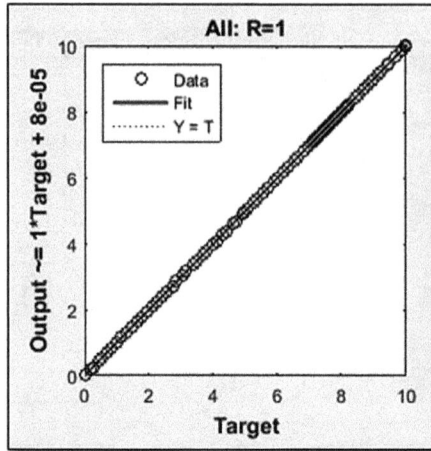

Figure 9.6 Combination with all sets of (training, testing, and validation) specific wear rate for Al-6061/SiC/Al$_2$O$_3$/REOs hybrid composites using Levenberg–Marquardt Algorithm.

their corresponding percentages of error were calculated and presented in Table 9.3.

9.4 CONCLUSIONS

This study presented a new model based on machine learning for predicting the wear rate of REO composites. The following conclusion have been drawn after the prediction of wear rate composites using the Levenberg–Marquardt Algorithm:

1. In this work, the back-propagation model system was created using the Levenberg–Marquardt Algorithm. The accuracy of the network after proper training has been assessed by way of evaluating the predicted value and experimental value of the specific wear rate of the REO-based composites. Results showed that the network 3-2-1-1 predicted results of specific wear rate are very comparable to the experimental results for Al-6061/SiC/Al$_2$O$_3$/REOs hybrid composites.
2. The correlation value (R) achieved is 0.98733 for all of the combined set of testing, training, and validation for the network formation 3-2-1-1 exclusive of error. The chosen network 3-2-1-1 generally showed the relationship of wear rate. It was observed that, at a very slight value of epochs (210), the network attains an accuracy of $4.3424e^{-10}$. The

closeness of predicted and experimental wear rate results suggested that ANN was proved to be better technique with a confidence level of 99%.

3. It is concluded that the Levenberg–Marquardt set of rules and the technique of ANN make for a superior device to evaluate the prediction values of wear rate of Al-6061/SiC/Al2O3/REOs hybrid composites. This will result in avoiding the material wastage for experimental tryout time and cost. The neural network method predicts the response more accurately in comparison to different techniques like the Taguchi RSM approach with a higher confidence level.

4. The main considerations that added to the diminished wear rates in composites after incorporation of REOs was located in the microstructural refinement, bringing about higher hardness and greater prominent ability of stress hardening with better ductility. The addition of cerium oxides resulted in a satisfactory length of the composite grains.

REFERENCES

1. Singh, H. and Bhowmick, H. Lubrication characteristics and wear mechanism mapping for hybrid aluminium metal matrix composite sliding under surfactant functionalized MWCNT oil. *Tribol Int* 2020; 145: 106152.

2. Rao, R.N., Das, S., Mondal, D.P., et al. Dry sliding wear maps for AA7010 (Al–Zn–Mg–Cu) aluminium matrix composite. *Tribol Int* 2013; 60: 77–82.

3. Ahlatci, H., Kocer, T., Candan, E., et al. Wear behaviour of Al/(Al2O3p+ SiCp) hybrid composites. *Tribol Int* 2006; 39: 213–220.

4. Yuan, L., Han, J., Liu, J., et al. Mechanical properties and tribological behavior of aluminum matrix composites reinforced with in-situ AlB2 particles. *Tribol Int* 2006; http://dx.doi.org/10.1016/j.triboint.2016.01.046

5. Rao, R.N., Das, S., Mondal, D.P., et al. Effect of heat treatment on the sliding wear behaviour of aluminium alloy (Al–Zn–Mg) hard particle composite. *Tribol Int* 2010; 43: 330–339.

6. Rao, R. and Das, S. Effect of matrix alloy and influence of SiC particle on the sliding wear characteristics of aluminium alloy composites. *Mater Des* 2010; 31: 1200–1207.

7. Tirth, V. Dry Sliding Wear Behavior of 2218 Al-Alloy-Al2O3 (TiO2) Hybrid composites, *Journal of Tribology* 2017; 140: 1–9.

8. Yadav, D. and Bauri, R. Effect of friction stir processing on microstructure and mechanical properties of aluminium. *Mater Sci Eng A* 2011; 528: 1326–1333.

9. Kumar, A., Patnaik, A. and Bhat, I.K. Investigation of nickel metal powder on tribological and mechanical properties of Al-7075 alloy composites for gear materials. *Powder Metallurgy* 2017; 60: 371–383.

10. Bai, M. and Xue, Q. Investigation of wear mechanism of SiC particulate-reinforced Al-20Si-3Cu-1Mg aluminium matrix composites under dry sliding and water lubrication. *Tribol Int* 1997; 30: 261–269.

11. Guo, X., Wang, L., Wang, M., et al. Effects of degree of deformation on the microstructure, mechanical properties & texture of hybrid-reinforced titanium matrix composites. *Acta Materialia* 2012; 60: 2656–2667.

12. Rohatgi, P.K., Schultz, B.F., Daoud, A., et al. Tribological performance of A206 aluminum alloy containing silica sand particles. *Tribol Int* 2010; 43: 455–466.

13. Sosimi, A.A., Gbenebor, O.P., Oyerinde, O., et al. Analysis of wear behavior of Al-CaCO3 composites using ANN and Sugeno-type fuzzy inference systems. *Neural Computing and Applications* 2020; 32: 13453–13464.

14. Zhang, Z., Friedrich, K. and Velten, K. Prediction on tribological properties of short fibre composites using artificial neural networks. *Wear* 2002; 252: 7–8; 668–675.

15. Gyurova, L.A., Justel, P.M. and Schlarb, A.K. Modeling the sliding wear and friction properties of polyphenylene sulfide composites using artificial neural networks. *Wear* 2010; 268: 5–6; 708–714.

16. Kumar, G.B.V. and R. Pramod. Artificial neural networks for predicting the tribological behaviour of Al7075-SiC metal matrix composites, *Journal of Materials Research and Technology*, Proceedings of the International Conference on Advances in Engineering and Technology (ICAET-2014), India (2014) 423–428 10.15224/978-1-63248-028-6-03-85.

17. D. Ozyurek, A. Kalyon, M.Yildrim, et al. Experimental investigations and prediction of wear properties of Al/SiC metal matrix composites produced by thixomoulding method using artificial neural networks, *Materials & Design* 2014;63:270–277.

18. K. Genel, S.C. Kurnaz and M. Durman. Modeling of tribological properties of alumina fiber reinforced zinc–aluminum composites using artificial neural network, *Mater Sci Eng A* 2003; 363: 203–210.

19. K. Shirvanimoghaddam, H. Khayyam, H. Abdizadeh, et al. Boron Carbide Reinforced Aluminium Matrix Composite: Physical, Mechanical Characterization and Mathematical Modelling, *Mater Sci Eng A* 2016; 21: 135–149.

20. P.S. Kumar, K. Manisekar and R. Narayanasamy. Experimental and prediction of abrasive wear behavior of sintered Cu-SiC composites containing graphite by using artificial neural networks, *Tribol Trans* 2014; 57: 455–471.

21. Borgaonkar, A.V. and Ismail, S. Tribological behavior prediction of composite MoS_2-TiO_2 coating using Taguchi coupled artificial neural network approach, *Proc IMechE, Part C: J Mechanical Engineering Science*, https://doi.org/10.1177/09544062211065995.

22. Bose, S. and Nandi, T. Statistical and experimental investigation using a novel multi-objective optimization algorithm on a novel titanium hybrid composite developed by lens process,*Proc IMechE, Part C: J Mechanical Engineering Science* 2020; 235 (16): 2911–2933.

23. Sharma, V.K., Kumar, V. and Joshi, R.S. Experimental analysis and characterization of SiC and RE oxides reinforced Al-6063 alloy based hybrid composites. *Int J Adv Manuf Technol* 108, 1173–1187 (2020). https://doi.org/10.1007/s00170-020-05228-7

24. Sharma, V.K., Kumar V. and Joshi, R.S. Investigation of rare earth particulate on tribological and mechanical properties of Al-6061 alloy composites for aerospace application. *Journal of Materials Research and Technology* 2019 Jul 1; 8(4): 3504–16.

25. Gyurova, L.A., Justel, P.M. and Schlarb, A.K. Modeling the sliding wear and friction properties of polyphenylene sulfide composites using artificial neural networks. *Wear*. 2010; 268: 5–6; 708–714.

26. Kumar, G.B.V. and R. Pramod. Artificial neural networks for predicting the tribological behaviour of Al7075-SiC metal matrix composites. *Journal of Materials Research and Technology*, Proceedings of the International Conference on Advances in Engineering and Technology (ICAET-2014), India (2014) 423–428 10.15224/978-1-63248-028-6-03-85.

27. D. Ozyurek, A. Kalyon, M. Yildrim, et al. Experimental investigations and prediction of wear properties of Al/SiC metal matrix composites produced by thixomoulding method using artificial neural networks, *Materials & Design* 2014; 63: 270–277.

Chapter 10

Systematic study of electrochemical machining processes for micromachining

Manpreet Singh and Sunil Kumar Paswan

10.1 INTRODUCTION

The classification of the microdrilling processes, as investigated in the electrochemical micromachining (ECMM) process, is a suitable ECMM arrangement is primarily comprised of a microtooling scheme: electrically powered system, mechanical machining rig, regulating system, and governed electrolytic flow structure to regulate electrochemical machining. As per the analysis, the most effective process parameters, such as machining voltage and electrolyte concentration, cause a higher material removal rate (MRR) with reduced overcut. The ECMM testing results and analysis will expand its application possibilities.

10.2 ELECTROCHEMICAL DRILLING (ECD)

Electrochemical drilling (ECD) is a regulated rapid electrolytic dissolution mechanism in which the workpiece functions as an anode (Figure 10.1). A thin distance separates the cathode instrument from the anode, through which an electrolyte flows [1–3]. The anodic substance dissolves locally when the current of electricity passes via electrolytic cell [4]. The electromagnetic electrolyte is commonly a distilled solution containing salt that

DOI: 10.1201/9781003360001-10

is pushed at increased pressure across the interelectrode gap to facilitate reaction materials, heat dissipation, and metal dissolution. The tubing instrument has a tubular look, consisting of brass, bronze, or stainless steel. The entire surface outside is normally isolated, except for the tip. NaCl, $NaNO_3$, $NaClO_3$, and their mixtures are four of the most often encountered electrolytes. There are mainly two key shortcomings of ECD: absence of tool encapsulation and stray elimination. The use of salt electrolytes causes clogging of the pores, which leads to insulation loss in the ECD. The stray elimination that happens on the holes internal side walls has a direct impact on process efficiency. It has been attempted to minimize stray exclusion by using high-quality insulation [5]. It was recently attempted with a dual pole tool. To decrease streaming on the hole surface, the two-pin tool used a metal bush outside the protected layer of a cathode tool. It was found that using a dual pole device reduces hole taper instead of an insulated method. This also increases the precision and consistency of the machining process. It is obvious that using a dual pole tool results in a reduced hole taper. When employing an insulated tip, however, the hole diameter difference along the hole depth must be 0.03 mm or less [6, 7].

10.3 ELECTROCHEMICAL MICROHOLE DRILLING

ECM was never employed for microhole drilling because the electric field was not confined, as well as because of the formation of the passive layer and the taper [8]. ECM has lately been used for pulsated current,

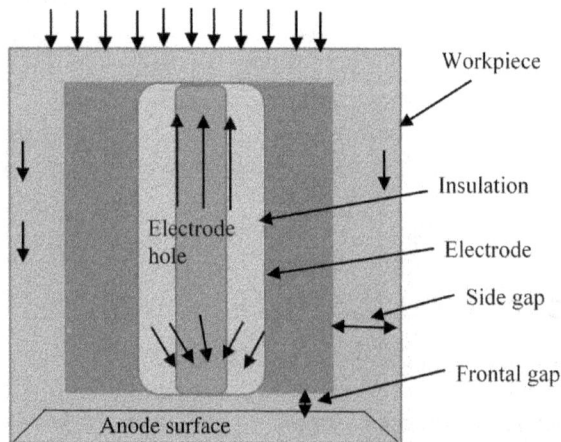

Figure 10.1 Working principle of electrochemical drilling (ECD) [1].

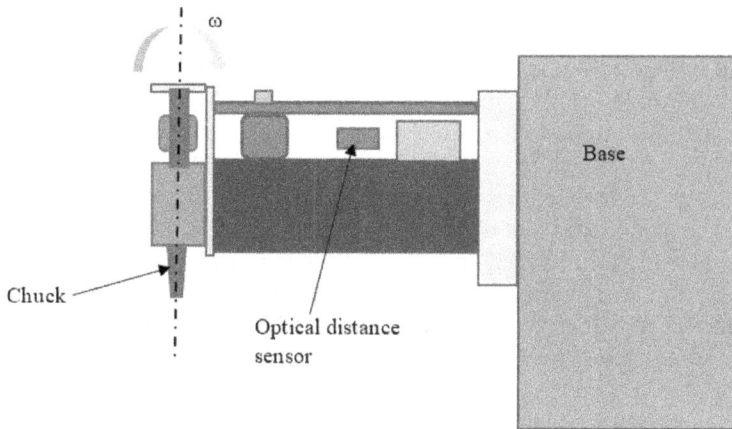

Figure 10.2 Electrochemical microhole drilling [8].

microdivergence regulator between cathode and anode, balanced electrode, and laterally separated instrumentation in microhole drilling [9]. Localized machining is closely aided by tool-side insulation (cathode) and microgap power. The anode-and-cathode-pulsated current helps to agitate the electrolyte and stimulate the electrochemical reaction. The machining process is difficult to continue because the dregs formed while performing machining may get stuck on the workpiece surface and the tool electrode [10]. However, these issues can be overcome if pulsed voltage is used. During the pulse offtime, the temperature of the electrolyte increases, and the dregs are washed away. Electrochemically cut microholes underneath the DC current and pulse voltage circumstances [11–13]. When the pulse current is utilized, the expanded section of the diameter of the hole relative to the electrode is significantly smaller than when DC current is used, as shown in Figure 10.2.

A profound microhole into the 304 SS was created with the use of a platinum balancing electrode (with half the surface area of the workpiece) and pulse voltage. The platinum balancing electrode was used to compensate between the electrolyte and two electrodes for the voltage fall differences [14]. The electrolyte-to-workpiece resistance is low, and the voltage drop is minimal [15] because the workpiece has a broader immersion zone than the tool. By generating a chromium oxide layer at the hole level, the electrode low voltage applied to the workpiece prevents further dissolution. The platinum balance electrode, according to the findings [16], avoids the development of a chrome oxide deposition over the hole surface when processing low-potential microholes.

10.4 SHAPED TUBE ELECTROLYTIC MACHINING (STEM)

Shaped tube electrolytic machining (STEM) (Figure 10.3) was industrialized to drill the holes of the workpiece having a large depth-to-diameter ratio that would be impossible to mine otherwise [17]. Although ECD was originally used to create accurate holes, its yields unsolvably hasten, impede, or limit the electrolyte flow channel. STEM is an improved ECD process in which the metal extracted is dissolved rather than precipitated with an electrolyte acid [18]. In STEM, acid electrolytes (sulfuric acid, nitric, as well as hydrochloric acid) with a concentration of 10-25% are favored. The mixture of neutral salt electrolytes (10%) and small acid electrolytes (1%) was used by the scholars to diminish the accumulation of sludge at interelectrode gap (IEG) in certain cases [19]. The ECM operating theory is followed in the STEM method. Regulated deplating of an electrically conductive substrate produces holes [20].

An electrolytic cell is created by an electrically conductive solution, which is negatively charged (cathode) metal electrode and a positively charged workpiece (anode). The cathode is mostly made up of an acid-resistant metal tube, such as titanium. Except for the tip [21], it is perfectly straightened and insulated throughout. The acid electrolyte is pushed through the tube to the tip under pressure, then returned to the top of the workpiece via a

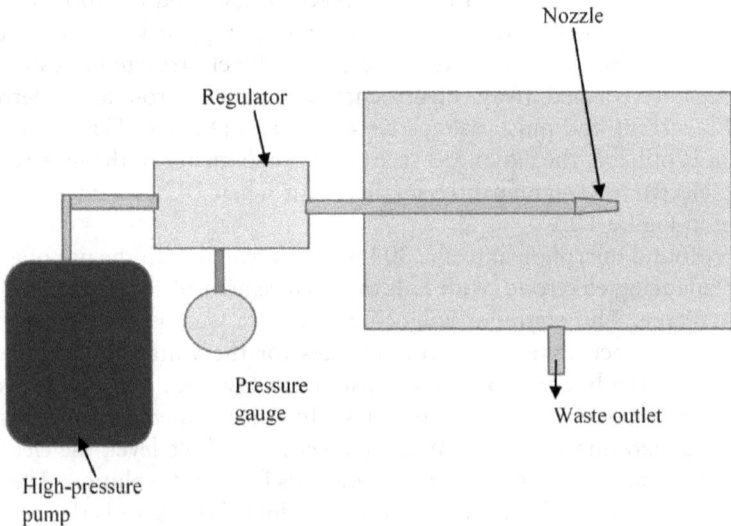

Figure 10.3 Shaped tube electrolytic machining (STEM) [17].

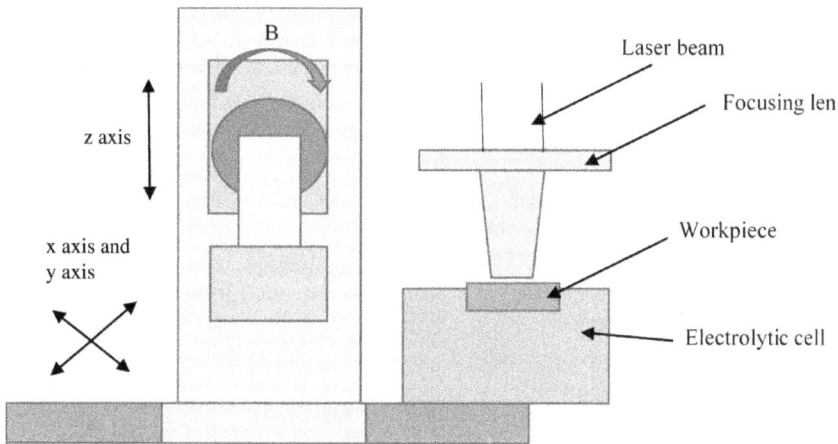

Figure 10.4 Hybrid laser STEM: (a) schematic diagram with the machining area and laser electrolyte jet coupling system, (b) Z-axis movement and coupling movement, and (c) hybrid tube electrode [24].

thin gap along the coated tubes outside as shown in the Figure 10.4 [22]. The electrode is fed at a steady rate that corresponds to the rate at which the workpiece substance dissolves [23]. (See Figure 10.4.)

STEM is ideal for drilling several holes of varying sizes or of the same dimension [24]. Grouped holes are generally parallel drilled, but they are boiled by means of guides that point the electrodes off the feed direction at compound angles. STEM (5V to15 V DC) has a lower operational voltage than a traditional electrochemical drilling (ECD) (10 to 30 V DC). The use of a typical ECD of conductive acid electrolytes rather than base electrolytes results in a lower voltage need [21, 22]. Since there is no mechanical interaction during STEM, the wall thickness is consistent in repeated development. The material molecule-by-molecule breakdown results in unstressed, high-integrity holes. STEM technology has been used to make microholes, holes with a high look-to-be proportion, large rectangular, elliptical, and holes with curved surfaces [23]. Careful monitoring of process parameters, complicated instrumentation, cutting-edge electrode processing, and the accessibility of complex CNC controls to regulate the STEM system operation made this STEM technique adaptation possible [25]. The existence of corrosive and dangerous acidic electrolytes, as well as the toxicity of gases emitted during the processing of a hole, necessitate the use of appropriate working procedures to protect the environment [26, 27].

10.5 ELECTROCHEMICAL JET MACHINING (ECJM)

Electrochemical jet machining (ECJM) refers to two methods for cutting micro- and macroholes and grooves with a pressured, charged acid electrolyte jet. ECJM is associated with processes such as CD, ESD, and JED [28]. (See Figure 10.5.)

These processes features and shortcomings are summarized in turn. Electrolyte jets are being investigated for new applications [29]. They are used for things like electron microscopic sample processing, micropart etching, semiconductor material polishing, and electrochemical micromachining [30, 31]. Owing to their poor aspect ratios, these applications were conducted with ECJM at a lower operating voltage. Consistent attempts are being carried out with a tight control of these systems machining parameters to increase the process capacity of ECJM (material removal rate and precision) [32].

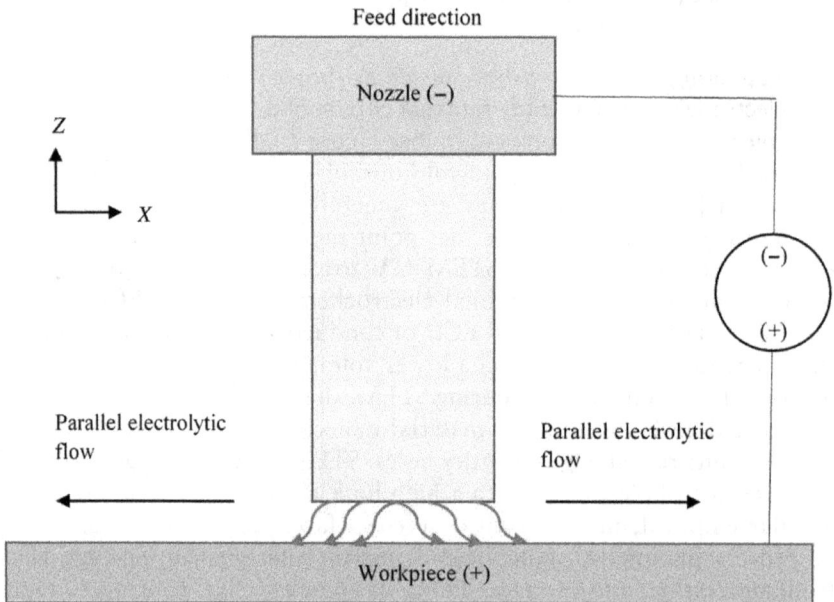

Figure 10.5 Unveiling of metal using electrochemical jet: (a) surfaces prepared on macroscale components, enabled by slot jet handing; (b) slot jet engraving generates equivalent electrolyte flow on pinch; (c) great surface areas prepared for more investigation, with grain disparity demarcated by the orientation-dependent microtopography [29].

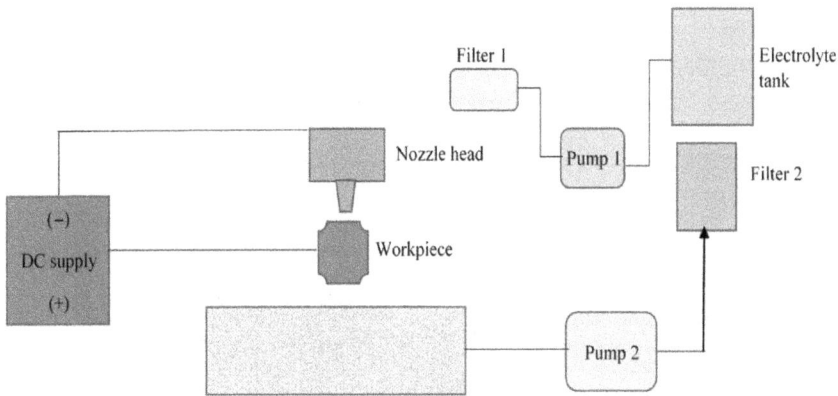

Figure 10.6 Schematic diagram of CD process [34].

10.6 CAPILLARY DRILLING (CD)

The capillary drilling (CD) method is used to dig a deep hole for STEM but not so deep that EDM cannot be drilled. The drill tube is a high-pressure (3bar to20 bar) glass capillary that allows electrolyte to flow [33]. The cathode is a platinum wire with a thin bore that matches that of the tube. To ensure that the electrolyte flow stability and orientation are not impaired at the tube tip, the wire is put roughly 2 mm back from the tip of the tube. To resolve the resistive route of a CD current, a higher operational voltage (100–200 V) is necessary due to the long route of the electrolyte flux [34]. This method involves drilling a trailing edge hole in gas turbine blades of high pressure (diameter 0.2mm to 0.5 mm, width 8 mm to16 mm). (See Figure 10.6.)

To address slight angle variations caused by the blade torso, the glass tube can be gently turned on a nose guide if necessary [35]. Boiling holes in output elements with locating and diameter acceptances of G (0.05 mm) is a very common use of this technique [36].

10.7 ELECTROSTREAM DRILLING (ESD)

The electrostream drilling (ESD) method for macro- as well as microholes generation involves a cost-effective and reliable nontraditional drilling process [37]. On the workpiece, a negatively charged flow of acidic electrolyte comes in touch with finely drawn glass tube dust [38]. (See Figure 10.7.)

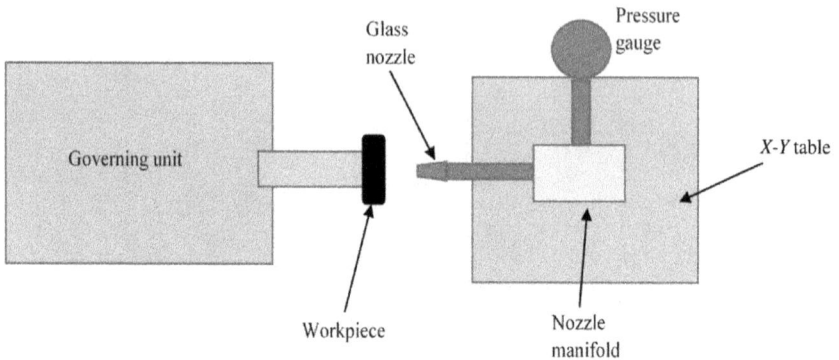

Figure 10.7 Experimental setup for EJD [41].

Acidic electrolyte (10% to 25% concentration) is injected into the glass tube nozzle at high pressure (3bar to 10 bar) [39–41]. The electrolyte jet works as a cathode as the platinum cable, which is threaded in one glass above the capillary and is linked to the negative terminal of DC power. The workpiece is used as an anode. Between the two electrodes, a suitable electrical capacity is applied [42]. Electrolytic dissolution removes the material from the workpiece as the electrolyte current impacts it [43]. The electrolyte flow then expels the metal ions and transports them away [44]. To maintain appropriate current flow, a longer and thinner electrolyte flow channel needs a higher voltage (150V to 850 V) [45–47]. Since there is a high chance of stray voltage when using high potential, it can be difficult to build an electrolyte device. Surface imperfections are caused by substrate homogeneity rather than method heterogeneity, as is seen often during ESD [48]. With a 3 mm thick electro jet drilling (EJD) of mild steel, the content removal rate is increased by up to 400 V with a glass pin (0.00025 m inner diameter) in the dwell feed method and decreases due to spark start-up [49]. When the voltage is increased, the current output reduces until it exceeds 500 V, after which point it improves [50]. A passivating coating on the hole top may be the source, which is broken by microsparking at higher voltages [51]. Based on Faraday law, the EJD model presented a theoretical calculation of the material removal rate (MRR) by evaluating a straight column of electrolyte between the instrument and the workpiece. This model is used to estimate how long it will take to machine an alloy [52].

10.8 JET ELECTROLYTIC DRILLING (JED)

Jet electrolytic drilling (JED) is a dwell drilling technique that eliminates the requirement for a nozzle to be inserted into the machined hole [53].

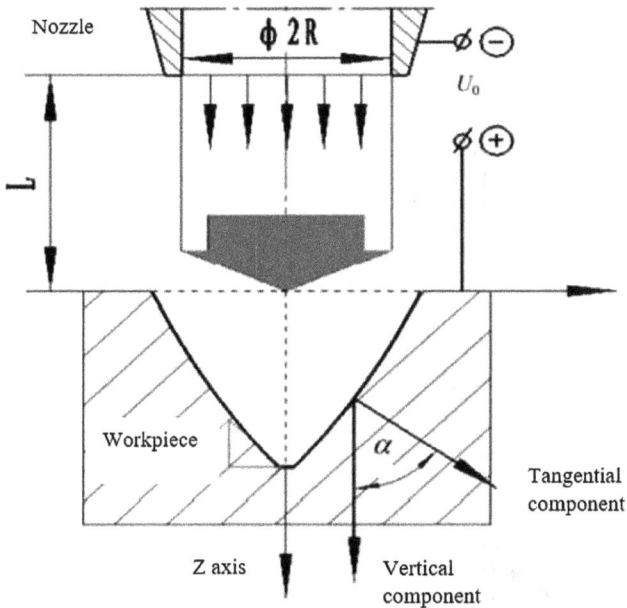

Figure 10.8 Schematic diagram of the jet electrolytic drilling technique [56]. A spacing of 2–4 mm must be preserved between the two electrodes. To attain the high current density required for significant stock removal, this technique requires a high operational voltage (400–800 V) and a high conductivity electrolyte [56].

To produce anodic dissolution of the workpiece material, a jet of electrolyte impinging on the workpiece at a pressure of 10 bar to 60 bar is utilized [54–56]. Despite the fact that the job component is anode, the electrolyte jet exits from the cathode procedure through this nozzle. The depth of the nozzle cavity, electrolyte friction, and overcut all have an effect on the lower bound of a hole that may be drilled. (See –0.8.)

10.9 LASER JET ECJM

The feasibility of the hybrid ECJM/laser beam method was demonstrated by research as a rapid means of precise micromachining [57–61]. Figure 10.9 shows the procedure. The electrolyte is injected into a jet cell, which is used as a standing jet that passes through a small pin at the workpiece (anode) [62]. A capillary tube is used to create the nozzle orifice, which has a diameter of 0.5 mm. A laser beam is steered into a central hole in the cathode, which is made of platinum [63–67]. A microcontroller controls the power supply (which is connected to the electrochemical cell), and the laser beams the on-off range. The current density in one experiment was up to 75 A/

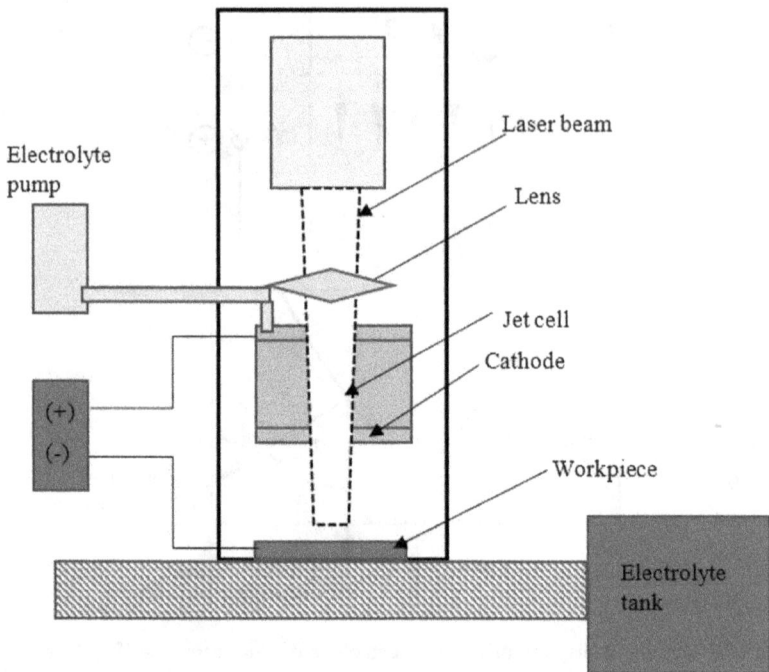

Figure 10.9 Experimental gadget: (a) drawing of experimental gadget and (b) realized experimental gadget [67].

cm². The electrolyte linear flow rate was kept at 10 m/s, and its spacing of the bowl anode was maintained at 3 mm [67]. An argon laser beam with a persistent intensity of 22 W was concentrated near the center of the jet by a 75 mm focus prism that passed through a beam expander [65]. This study demonstrated that a laser jet ECJM can rapidly eat microholes in direct-to-metal (DTM) painted metals [67]. Furthermore, it was discovered that using a laser jet greatly reduced overcutting [64]. At the current density of 0.6 A/cm², the chloride solution removes nearly the same volume of material from microholes cut with and without a laser jet. However, instead of a hole depth of 0.011 mm, a deeper depth of 0.055 mm was achieved by using the laser jet [67].

The efficacy of the laser jet ECJM has been calculated by comparing the volume of material collected to the hole depth (v/d). It is also been used to evaluate outstanding performance [68]. The lower the v/d figure, the less stray and excessive current there is. The approximate v/d values in the occurrence and nonappearance of the laser beam depend on the current density. When using a laser jet, the v/d values are smaller, and the effect on

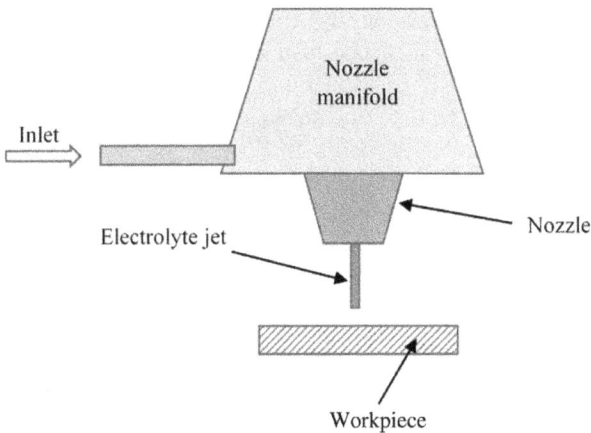

Figure 10.10 Experimental setup of the laser jet ECJM [70].

steel is greater than on nickel [69]. A neutral solution, according to research, may be employed to do high-speed microdrilling on a range of metals and alloys. The effects of a YAG-pulsed laser beam on the form factor were found to be negligible in electrochemical jet grading using a 150 mm thick nickel foil and a nonpassifying medium (sodium chloride). The factor form is proportionate to the volume of a machined hole of the perfect hole (diameter D). For cylindrical geometry holes, form factors have been determined [70]. (See Figure 10.10.)

The study proposed seeks to improve the efficiency of the jet-ECMD mechanism by growing the MRR and dropping the hole taper. In order to get the consequence of acoustic electrolyte handling, an experimental setup was built in this study to transmit ultrasonic vibrations through the electrolyte jet into the machining zone. In addition, the setup included a link to the coaxial air flux as well as an electrolyte jet to essence the current density in the machining field [71]. The influence of various factors such as voltage (direct current and pulsed direct current), electrolyte concentration, electrical strain, and working gap on the jet-ECMD process responses such as MRR and hole taper of ultrasonic vibrations was studied. As a result, four jet-ECMD outcomes options were evaluated. Therefore, four jet-ECMD alternatives for results were compared.

10.10 CONCLUSIONS

Recent technological advancements have allowed the ECM to be used in the drilling of circular microholes. Recent advancements in the ECM process have provided it with numerous advantages, such as its machining rates have become flexible and can be adjusted with control of the electric current, it generally uses environmentally friendly nontoxic and noncorrosive electrolytes, thermal and/or mechanical residue stresses are not generated in the workpiece material by this machining process, and the workpiece surface property does not change during the machining process.

REFERENCES

1. Zhu, D., Wang, W., Fang, X. L., Qu, N. S., & Xu, Z. Y. 2010. Electrochemical drilling of multiple holes with electrolyte-extraction. *Cirp Annals, 59*(1), 239–242.
2. Wei, C., Xu, K., Ni. J., Brzezinski, A. J., Hu, D. 2011. A finite element-based model for electrochemical discharge machining in discharge regime, *International Journal Advanced Manufacturing Technology, 54,* 987–995.
3. Barker, T.B. 1986. Quality engineering by design: Taguchi's Philosophy, *Quality Progress,* December, 33–42.
4. Katsushi Furutani and Hideaki Maeda 2008. Machining a glass rod with a lathe-type electrochemical discharge machine, *Journal of Micro Mechanics and Micro Engineering, 18,* 065006 (8pp).
5. Barker T.B. 1990. Engineering quality by design, *Marcel Dekker Inc,* New York.
6. Furutani, K. and Maeda, H. 2008. Machining a glass rod with a lathe-type electro-chemical discharge machine, *Journal of Micromechanics & Microengg., 18,* 065006 (8pp).
7. Bhattacharyya, B., Doloi, B., Sorkhel, S.K. 1999. Experimental investigations into electrochemical discharge machining (ECDM) of non-conductive ceramic materials, *Journal of Materials Processing Technology, 95,* 145–154.
8. Jalali, M., Maillard, P., & Wüthrich, R. 2009. Toward a better understanding of glass gravity-feed micro-hole drilling with electrochemical discharges. *Journal of Micromechanics and Microengineering, 19*(4), 045001.
9. Chak, S. K. and Rao, P. V. 2008. The drilling of Al2O3 using a pulsed DC supply with a rotary abrasive electrode by the electrochemical discharge process, *International Journal Advanced Manufacturing Technology, 39,* 633–641.
10. Chak., P. V. Rao. 2012. Machining of SiC by ECDM Process using Different Electrode Configurations under the Effect of Pulsed DC, *All India Manufacturing Technology, Design and Research Conference,* Vol. 1, 513–519.
11. Cheng C. P., Wu, K. L, Mai, C. C., Yang, C. K., Hsu, Y. S., Yan, B. H. 2010. Study of gas film quality in electro-chemical discharge machining, *International Journal of Machine Tools & Manufacture, 50,* 689–697, 2010.
12. Cheng, C. P., Wu, K. L., Mai, C. C., Hsu, Y. S. and Yan, B. H. 2010. Magnetic field-assisted electrochemical discharge machining, *Journal of Micromechanics & Micro Engineering, 20,* 075019 (7pp).
13. Cook., G.B. Foote., P. Jordan., B.N. Kalyani. 1973. Experimental studies in electro-machining, *Trans. ASME, Journal of Engineering for Industry* pp. 945–950.

14. Xiaowel., J. Zhixin., Z. Jiaqi., L. Jinchun. 1997. A combined machining process for the production of a flexure hinge, *Journal of Material Processing Technology*, 71, 373–376.

15. Yan, B. H., Yang, C. T., Huang, F. Y. and Lu, Z. H. 2007. Electrophoretic deposition grinding (EPDG) for improving the precision of micro holes drilled via ECDM, *Journal of Micromechanics & Micro engineering*, 17, 376–383.

16. Yang, C. K., Wu, K. L., Hung, J. C., Lee, S. M., Lin, J. C., Yan, B. H. 2011. Enhancement of ECDM efficiency and accuracy by spherical tool electrode, *International Journal of Machine Tools & Manufacture*, 51, 528–535.

17. Glebov, V. V., Danilenko, I. N., & Ratushinsky, R. I. 2018. Hole drilling and milling of magnetic alloys parts by shaped tube electrolytic machining. In *MATEC Web of Conferences* (Vol. 226, p. 03017). EDP Sciences.

18. Coteață, M., Ciofu, C., Slătineanu, L., Munteanu, A. and Dodun, O. 2009. Establishing the electrical discharge weight in electrochemical discharge drilling, *Int J Mater Form*, Vol. 2 Suppl 1: 673–676.

19. Basak, I. and Ghosh, A. 1996. Mechanism of spark generation during electrochemical discharge machining: a theoretical model and experimental verification, *Journal of Materials Processing Technology*, 62, 46–53.

20. Ghosh, A. 1997. Electro chemical discharge machining: principle & possibilities, *Sadhana*, 22(3), 435–447.

21. Han, M. S., Min, B. K. and Lee, S. J. 2009. Geometric improvement of electrochemical discharge micro-drilling using an ultrasonic-vibrated electrolyte, *Journal of Micromechanics & Micro Engineering*, 19, 065004 (8pp).

22. Han, M. S., Min, B. K., Lee, S. J. 2007. Improvement of surface integrity of electro-chemical discharge machining process using powder-mixed electrolyte, *Journal of Materials Processing Technology*, 191, 224–227.

23. Jain, V. K. and Adhikary, S. 2008. On the mechanism of material removal in electrochemical spark machining of quartz under different polarity conditions, *Journal of Materials Processing Technology*, 200, 460–470.

24. Wang, Y., & Zhang, W. 2021. Theoretical and experimental study on hybrid laser and shaped tube electrochemical machining (Laser-STEM) process, *The International Journal of Advanced Manufacturing Technology*, 112(5), 1601–1615.

25. Jain, V.K., Dixit, P.M., Pandey, P.M. 1999. On the analysis of the electrochemical spark machining process, *International Journal of Machine Tools & Manufacture*, 39, 165–186.

26. Yang, C.-K., Cheng, C.-P., Mai, C.-C., Wang, A. C., Hung, J.-C., Yan, B.-H. 2010. Effect of surface roughness of tool electrode materials in ECDM performance, *International Journal of Machine Tools & Manufacture*, 50, 1088–1096.

27. Zheng, Z. P., Su, H. C., Huang, F. Y. and Yan, B. H. 2007. The tool geometrical shape and pulse- off time of pulse voltage effects in a Pyrex glass electrochemical discharge micro drilling process, *Journal of Micromechanics & Micro Engg.*, 17, 265–272.

28. Ziki, J.D., Didar, T.F, Wuthrich, R. 2010. Micro-texturing channel surfaces on glass with spark assisted chemical engraving, *International Journal of Machine Tools & Manufacture*, 57, 66–72.

29. Speidel, A., Xu, D., Bisterov, I., Mitchell-Smith, J., & Clare, A. T. 2021. Unveiling surfaces for advanced materials characterisation with large-area electrochemical jet machining, *Materials & Design*, 202, 109539.
30. Jana., A. Ziki., T. F. Didar., R. Wuthrich. 2012. Micro-texturing channel surfaces on glass with spark assisted chemical engraving, *International Journal of Machine Tools & Manufacture*, 57, 66–72.
31. Jawalkar., A. K. Sharma., P. Kumar. 2012. Micromachining with ECDM: research potentials and experimental investigations, *World Academy of Science, Engineering and Technology*, 61.
32. Jawalkar., P. Kumar., A. K. Sharma. 2012. On mechanism of material removal and parametric influence while machining soda lime glass using Electro-Chemical Discharge Machining (ECDM), *All India Manufacturing Technology, Design and Research Conference*, Vol. 1, 440–446.
33. Jiang, B., Lan, S., Ni, J., Zhang, Z. 2014. Experimental investigation of spark generation in electrochemical discharge machining of non-conducting materials, *Journal of Materials Processing Technology*, 214, 892–898.
34. De Silva, A. K., Gamage, J. R., & Harrison, C. S. 2017. Assessment of environmental performance of shaped tube electrolytic machining (STEM) and capillary drilling (CD) of superalloys, *CIRP Annals*, 66(1), 57–60.
35. Bhondwe, K. L., Yadava, V., Kathiresan, G. 2006. Finite element prediction of material removal rate due to electro-chemical sparks machining, *International Journal of Machine Tools & Manufacture*, 46, 1699–1706.
36. Khas., A. Manna. 2010. A study on MECD-machining during drilling of electrically non-conductive ceramic, *National Conference on Advancements and Futuristic Trends in Mechanical and Materials Engineering*, 61, 19–20.
37. Kim, D.J., Ahn, Y., Lee, S.H. and Kim, Y.K. 2006. Voltage pulse frequency and duty ratio effects in an electrochemical discharge micro-drilling process of Pyrex glass, *International Journal of Machine Tools & Manufacture*, 46, 1064–1067.
38. Kozak., K. E. Oczos. 2001. Selected problems of abrasive hybrid machining, *Journal of Materials Processing Technology*, 109, 360–366.
39. Kulkarni, A., Sharan, R., Lal, G. K. 2002. An experimental study of discharge mechanism in electro-chemical discharge machining, *International Journal of Machine Tools & Manufacture*, 42, 1121–1127.
40. Kulkarni, A. 2003. Measurement of temperature transients in electrochemical discharge machining process, *Journal of Materials Processing Technology*, 200, 460–470.
41. Sen, M., & Shan, H. S. 2007. Electro jet drilling using hybrid NNGA approach, *Robotics and Computer-Integrated Manufacturing*, 23(1), 17–24.
42. Kurafuji., K. Suda. 1968. Electrical discharge drilling of glass, *Annals of the CIRP*, 16, 415.
43. Lijo., S. Somashekhar., H. Ranganayakulu and J. Ranganayakulu. 2012. Experimental investigation and response surface modeling of metal removal rate in electrochemical discharge machining, *All India Manufacturing Technology, Design and Research Conference*, Vol. 1, 499–504.
44. Liu, J. W., Yue, T. M. and Guo, Z. N. 2013. Grinding-aided electrochemical discharge machining of particulate reinforced metal matrix composites, *International Journal Advanced Manufacturing Technology*, 8, 4846.

45. Liu, J. W., Yue, T. M., Guo, Z. N. 2010. An analysis of the discharge mechanism in electro chemical discharge machining of particulate reinforced metal matrix composites, *International Journal of Machine Tools & Manufacture*, 50, 86–96.

46. Liu., T. M. Yue., Z. N. Guo. 2009. Wire electrochemical discharge machining of Al2O3 particle reinforced aluminum alloy 6061, *Materials and Manufacturing Process*, 24, 446–453.

47. Mallick., B. Doloi., B. Bhattacharyya., B.R. Sarkar. 2012. Investigations into Travelling Wire Electrochemical Discharge Machining Process, *All India Manufacturing Technology, Design and Research Conference*, Vol. 1, 480–486.

48. Mediliyegedara, T.K.K.R., De Silva, A.K.M., Harrison, D.K., McGeough, J.A. 2005. New developments in the process control of the hybrid electro chemical discharge machining (ECDM) process, *Journal of Materials Processing Technology*, 167, 338–343.

49. Mochimaru, M. Ota, K Yamaguchi. 2012. Micro hole processing using electro-chemical discharge machining, *Journal of Advanced Mechanical Design, Systems and Manufacturing*, Vol. 6, No. 6.

50. Peace G.S. 1993. Taguchi Methods: A hands on approach, *Addison Wesley*, New York.

51. Peng, W.Y. and Liao, Y.S. 2004. Study of electrochemical discharge machining technology for slicing non-conductive brittle materials, *Journal of Materials Processing Technology*, 149, 363–369.

52. Raghuram., T. Pramila., Y. G. Srinivasa., K. Narayanasamy. 1995. Effect of the circuit parameters on the electrolytes in the electrochemical discharge phenomenon, *Journal of Materials Processing Technology*, 52, 301–318.

53. Ross P.J. 1988. Taguchi techniques for quality engineering, McGraw-Hill Book Company, New York.

54. Roy R.K., 1990, A primer on Taguchi method, Van Nostrand Reinhold, New York.

55. Han, M.S., Min, B.K., Sang, S.J. 2011. Micro-electrochemical discharge cutting of glass using a surface-textured tool, *Annals of the CIRP Journal of Manufacturing Science and Technology*, 4,362–369.

56. Zhang, H., Xu, J. W., & Zhao, J. S. 2011. Modeling and Experimental Investigation of Jet Electrolytic Drilling. In *Key Engineering Materials* (Vol. 458, pp. 277–282). Trans Tech Publications Ltd.

57. Huang, S. F., Liu, Y., Li, J., Hu, H. X., Sun, L. Y. 2014. Electrochemical discharge machining micro-hole in stainless steel with tool electrode high-speed rotating, *Materials and Manufacturing Processes*, 29, 634–637.

58. Sankar, A. R., Bindu, V.S.S, Das, S. 2011, Coupled effects of gold electroplating and electrochemical discharge machining processes on the performance improvement of a capacitive accelerometer, *Microsystems Technology*, 17, 1661–1670.

59. Sarkar, B.R., Doloi, B., Bhattacharyya, B. 2006. Parametric analysis on electrochemical discharge machining of silicon nitride ceramic, *International Journal Advanced Manufacturing Technology*, 28, 873–881.

60. Silva., J.A. McGeough. 1998. Process monitoring of electrochemical micromachining, *Journal of Materials processing Technology*, 76, 165–169.

61. Biswas., B. R. Sarkar., B. Doloi and B. Bhattacharyya. 2012. Parametric Optimization of μ-ECDMing of Silicon Nitride Ceramics, All India Manufacturing Technology, *Design and Research Conference*, Vol. 2, 1079–1084.

62. Byrne D.M. and Taguchi S. 1987. The Taguchi approach to parameter design, *Quality Progress*, 19–26.

63. Cao, X. D., Kim, B. Y. and Chu, C. N. 2013. Hybrid micromachining of glass using ECDM and micro grinding, *International Journal of Precision Engineering and Manufacturing*, 14(1), 5–10.

64. Skrabalak, G., Zybura, M, Skrabalak, Ruszaj A. 2004. Building of rules base for fuzzy-logic control of the ECDM process, *Journal of Materials Processing Technology*, 149, 530–535.

65. Tandon., V. K. Jain., P. Kumar., K. P. Rajurkar. 1990, Investigations into machining of composites, *International Journal of Precision Engineering*, 227–238.

66. Tsuchiya., T. Inoue., M. Miyazaiki. 1985. wire electro-chemical discharge machining of glasses and ceramics, *Bullt. Japan Soc. of Prec. Engg.*, 19, 73.

67. Tang, Y. 2002. Laser enhanced electrochemical machining process, *Materials and Manufacturing Processes*, 17(6), 789–796.

68. Wei, C., Hu, D., Xu, K., Ni, J. 2011. Electro-chemical discharge dressing of metal bond micro-grinding tools, *Inter. Journal of Machine Tools & Manuf.*, 51, 165–168.

69. West, J. and Jadhav, A. 2007. ECDM methods for fluidic interfacing through thin glass substrates and the formation of spherical micro cavities, *Journal of Micromechanics & Micro Engineering*, 17, 403–409.

70. Wuthrich, R. and Fascio, V. 2005 Machining of non-conducting materials using electrochemical discharge phenomenon—an overview, *International Journal of Machine Tools & Manufacture*, 45, 1095–1108.

71. Wüthrich, R. and Allagui, A. 2010. Building micro and nano systems with electrochemical discharges, *Electrochimica Acta*, 55, 8189–8196.

Chapter 11

Study of mechanical and microstructural properties of titanium chips fabricated by large strain machining process

Deepak Sharma, Kunal Arora, Vinod Kumar, and Vipin Kumar Sharma

11.1 INTRODUCTION

For the past two decades, the use of nanostructured materials in the fields of aerospace, defense, aviation, and manufacturing has increased. Titanium and other lightweight alloys have been extensively used in these fields [1]. Since these alloys also possess great strength, they have been widely used in these fields. Nanostructured material possesses excellent mechanical, physical, and chemical properties; therefore their demand has gained attention in the near past. So research and development need to be carried out across globe for the development of nanostructured material [2, 3]. The grain size of nanostructured material varies from 1 to 10 nm (nanometers), they are different from their coarse grain counterparts. Nanostructured material possesses significant mechanical properties such as high strength, low ductility, and higher superplasticity at lower temperatures as compared to their coarse grain counterparts. These extraordinary properties of nanostructured materials are due to smaller grain size compared to that of their coarse grain counterparts [4, 5]. Nanostructured materials are divided into three categories. In the first category are materials that have reduced dimension

DOI: 10.1201/9781003360001-11

in the form of nanometer-sized particles. Inert gas condensation, various aerosol techniques are a few methods to generate this class of material. The second category includes materials that have a nanometer-sized microstructure limited to a thin surface region of the material. Laser beam and ion implantation are a few techniques to modify this class of nanostructured material [6, 7]. The third category is actually a subgroup of the second category in which the surface region is structured by a structural pattern on the free surface. This type of nanostructured material can be synthesized by lithography.

Mainly two categories of approach are used to synthesize nanostructured materials: a bottom-up approach and a top-down approach [8–12]. However, there are a few limitations of these approaches. One such limitation is that, in order to create large plastic strain, multiple stages of deformation are required, and it is found to be difficult to deform metals and alloys of high strength due to different constraints applied by operating tools. In a few cases, preheating at elevated temperatures is mandatory, not all materials can be prepared by these methods, and the tools and equipment are very costly [13].

Therefore, in the past decade, the machining-based deformation process has emerged as a feasible alternative to conventional severe plastic deformation methods. By imposing a large uniform plastic strain in a single pass of a cutting tool, the chip formation resulting from it offers an alternative to produce nanostructured and UFG materials. Shear strain can vary from a range of 1–15 in various materials [14–17]. But unlike in conventional methods, control over shape and dimension of fine-grained chips is limited. But microstructure refinement can be done using large strain machining.

11.2 MATERIALS AND METHODS

11.2.1 Material selection

Since a lot of work has been done on materials that are easy to machine, so material that is hard to fabricate was taken into consideration in the present work. Titanium alloy (Ti6Al4V), which is hard to machine, was taken as a material for the study. Ti6Al4V was used for following reasons. This alloy has high tensile strength and toughness. It is light in weight and has extraordinary corrosion resistance properties. This grade of titanium alloy withstands extreme temperature limits and is widely used in the defense, aviation, spacecraft, and biomedical fields. The chemical composition and properties are reported in Tables 11.1 and 11.2, respectively [18, 19].

Table 11.1 Composition of Ti6Al4V

Alloy	Ti	Al	V	Fe	O	C	N	H
Ti6Al4V	87.6–91%	5.5–6.75%	3.5–4.5%	≤0.4%	≤0.02%	≤0.08%	≤0.05%	≤0.015%

Table 11.2 Properties of Ti6Al4V

S. No.	Parameter	Value
1.	Density	4.429–4.512 mg/m³
2.	Ductility	0.05–0.18
3.	Hardness	3370–3730 Mpa
4.	Melting temperature	1178–1933 K

11.2.2 Experimental setup

In the current investigation, the effects of various large strain machining process parameters are studied. The process parameters are tool rake angle, feed, and speed. Titanium alloy is used as a raw material. A 2hp universal lathe with rotary configuration is used to produce the strips. Specially designed tools made with carbide tips were used to replace conventional cutting tools, and a specially designed tool post was also prepared. The machining was done radially at different speeds and feeds. Machining was done without using lubricant. With the continuous feeding of the tool, strips of different lengths were fabricated from the tool. Experiments were conducted as per Box–Behnken design (BBD) with the help of a Design-Expert.

11.3 DESIGN OF EXPERIMENT

In this chapter, we develop the design for the surface roughness and microhardness of Ti6Al4Vchips fabricated from large strain machining. The impact of various input parameters on responses has been studied after producing the chips. The various input parameters were rake angle, feed, and speed. Experiments were designed by using Design-Expert software. This software runs the experiments randomly. Box–Behnken design technique was used for conducting the experiment. It plays an important role for achieving the high efficiency of the response surface methodology. Box–Behnken design experiments require fewer experiments as compared to Central Composite design. The proposed Box–Behnken design requires 17 experiments since there are three parameters.

Table 11.3 Selected Factors and their ranges

Factors	Level 1	Level 2	Level 3
Tool rake angle (°)	−5	0	5
Speed(m/min)	40	80	120
Feed (mm)	0.05	0.10	0.15

Table 11.4 Trail condition using BBD

STD	Run	A:Rake(°)	B:Speed(rpm)	C:Feed(mm)
10	1	0	720	0.05
12	2	0	720	0.15
4	4	5	720	0.10
1	5	−5	420	0.10
3	6	−5	720	0.10
8	7	5	570	0.15
14	8	0	570	0.10
5	9	−5	570	0.05
15	10	0	420	0.10
11	11	0	420	0.15
17	12	0	570	0.10
13	13	0	570	0.10
7	14	−5	570	0.15
16	15	0	570	0.10
9	16	0	420	0.05
2	17	5	420	0.10

11.3.1 Factors selection

The process parameters with their different levels have been selected after exhaustive literature survey. The selected parameters are shown in Table 11.3.

11.3.2 Box–Behnken design

Response surface methodology (RSM) is used for achieving the high efficiency of the response surface methodology. RSM is a statistical technique that is used to maximize the output parameters. This technique is used in developing new technical studies. This technique is very cost-effective. The efficiency of the Box–Behnken design is higher for an experiment involving three factors and three levels. The proposed Box–Behnken design has 17 experiments, as depicted in Table 11.4. The experiments in software were run randomly. Design-Expert software was used for determining 2D and 3D plots. These graphs give a clear idea of the effect of process variables over other variables. Further, the model was verified by performing experiments, taking two sets of random input values. The response values obtained through experiments were similar with equation values using the model.

11.4 RESULTS AND DISCUSSIONS

The result of microhardness and surface roughness for Ti6Al4V is shown in Table 11.5 and Figure 11.1.

The effects of various process parameters, such as rake angle, feed, and speed, on mechanical properties such as microhardness and surface roughness were analyzed, as well as the effect of shear strain and surface morphology. A Ti6Al4V titanium alloy cylindrical rod of 60mm diameter was used for this experimental work. Design-Expert software was used for the experimental work. In this proposed investigation, all three factors were varied at three levels. After conducting the 17 trials, the mean value of all the factors were tabulated.

Table 11.5 Experimental obtained values of microhardness and surface roughness of Ti6Al4V chips

STD	Run	A:Rake(^{0})	B:Speed (rpm)	C:Feed(mm)	Microhardness (HV)	Surface roughness(μm)
10	1	0	720	0.05	374.49	0.36
12	2	0	720	0.15	374.93	0.32
6	3	5	570	0.05	393.13	0.48
4	4	5	720	0.10	372.14	0.38
1	5	-5	420	0.10	416.62	0.42
3	6	-5	720	0.10	377.66	0.46
8	7	5	570	0.15	394.87	0.56
14	8	0	570	0.10	397.98	0.39
5	9	-5	570	0.05	402.09	0.63
15	10	0	570	0.10	413.68	0.38
11	11	0	420	0.15	414.96	0.31
17	12	0	570	0.10	398.24	0.37
13	13	0	570	0.10	398.99	0.40
7	14	-5	570	0.15	401.88	0.54
16	15	0	570	0.10	400.37	0.41
9	16	0	420	0.05	413.12	0.44
2	17	5	420	0.10	410.71	0.37

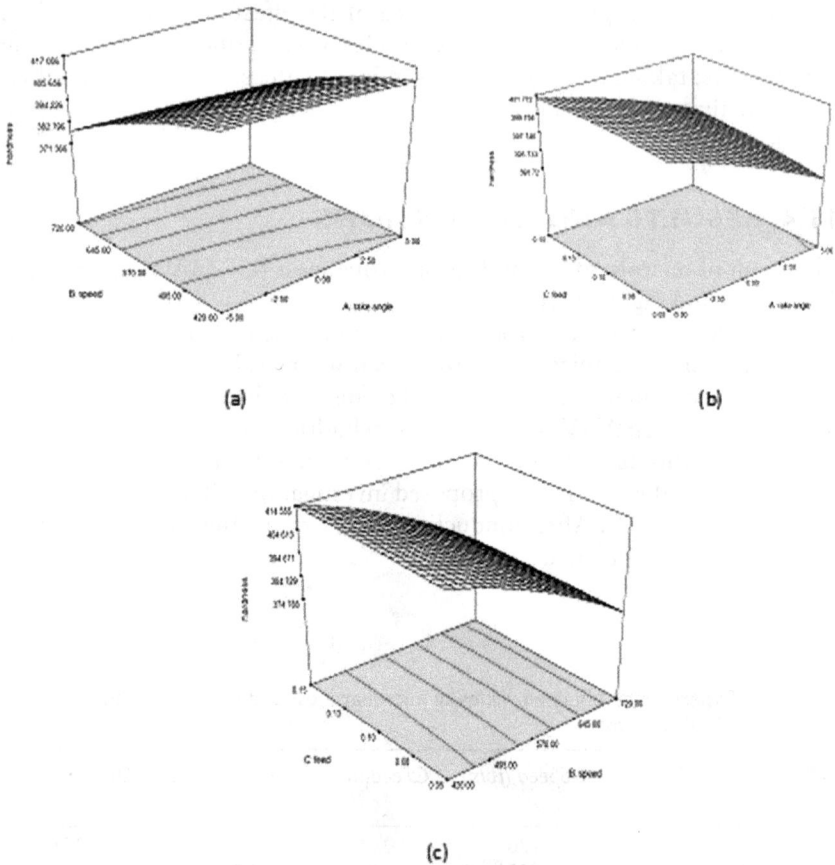

(a)

(b)

(c)

Figure 11.1 3D plot of variation of (a) speed and rake vs. microhardness, (b) feed and rake vs. microhardness, (c) feed and speed vs. microhardness.

11.4.1 Analysis for microhardness

The micro hardness of the strips was measured by a Vicker hardness tester machine. The equation developed for response hardness is

$$\text{Microhardness} = 398.84 - 3.42 \times A - 19.44 \times B + 0.44 \times C - 0.49 \times A \times A - 4.17 \times B \times B - 0.35 \times C \times C + 0.04 \times A \times B + 0.43 \times A \times C - 0.35 \times B \times C \quad (11.1)$$

As we can see in Table 11.6, B contributed maximum to the model with a contribution of 95.02%. So it is the most significant factor in the model. There is a 60.98% chance that a "Lack of Fit–F-value" this large could

Table 11.6 ANOVA table for microhardness

Source	Sum of squares	DF	Mean square	F-value	Prob>F	Contribution (%)	Status
Model	3488.60	9	387.62	387.85	<0.0001		Significant
A	93.43	1	93.43	93.49	<0.0001	2.66	Significant
B	3326.42	1	3326.42	3328.36	<0.0001	95.02	Significant
C	1.56	1	1.56	1.56	0.2520	0.044	Not significant
A^2	0.99	1	0.99	0.99	0.3528	0.02	Significant
B^2	72.92	1	72.92	72.96	<0.0001	2.08	Significant
C_2	0.51	1	0.51	0.51	0.4985	0.014	Not significant
A_B	0.00722	1	0.00722	0.007	0.9346	0.0002	Not significant
A_C	0.73	1	0.73	0.73	0.4207	0.020	Not significant
B_C	0.49	1	0.49	0.49	0.5064	0.013	Not significant
Residual	7.00	7	1.00				Not significant
Lack of fit	3.55	3	0.89	0.77	0.2043		Not significant
Pure error	3.45	4	1.15			0.098	
Cor total	3495.59	16					

Significant factor: A, B, B^2

Insignificant factor: A^2, C, C^2, AC, BC, AB

Table 11.7 Different ANOVA parameters

R-squared	0.9980
Adj R-squared	0.9954
Pred R-squared	0.9847
Adeq precision	59.628

not occur due to noise. Nonsignificant lack of fit is good – we want the model to fit. The values of R-squared given by the software are shown in Table 7.

The effects of rake angle and speed on hardness at different process parameters are shown in Figure 11.2. The trend portrays a reduction in hardness with the increase in cutting speed (from 79 to 136 m/min) and the decrease in rake angle (from +5 to –5°). With the decrease in rake angle, the forces and stresses in the PDZ increase, causing a driving force to the dislocations piled up along the grain boundaries, as well as low-angle grain boundaries, resulting in subgrain formation or grain refinement. This grain refinement increases the dislocation density, causing increased hardness of the material.

For a given rake angle, as the cutting speed increases, the temperature in the PDZ increases, thereby promoting grain growth in the material, because of which the hardness of the material decreases. Dinakar Sagapuram et al.

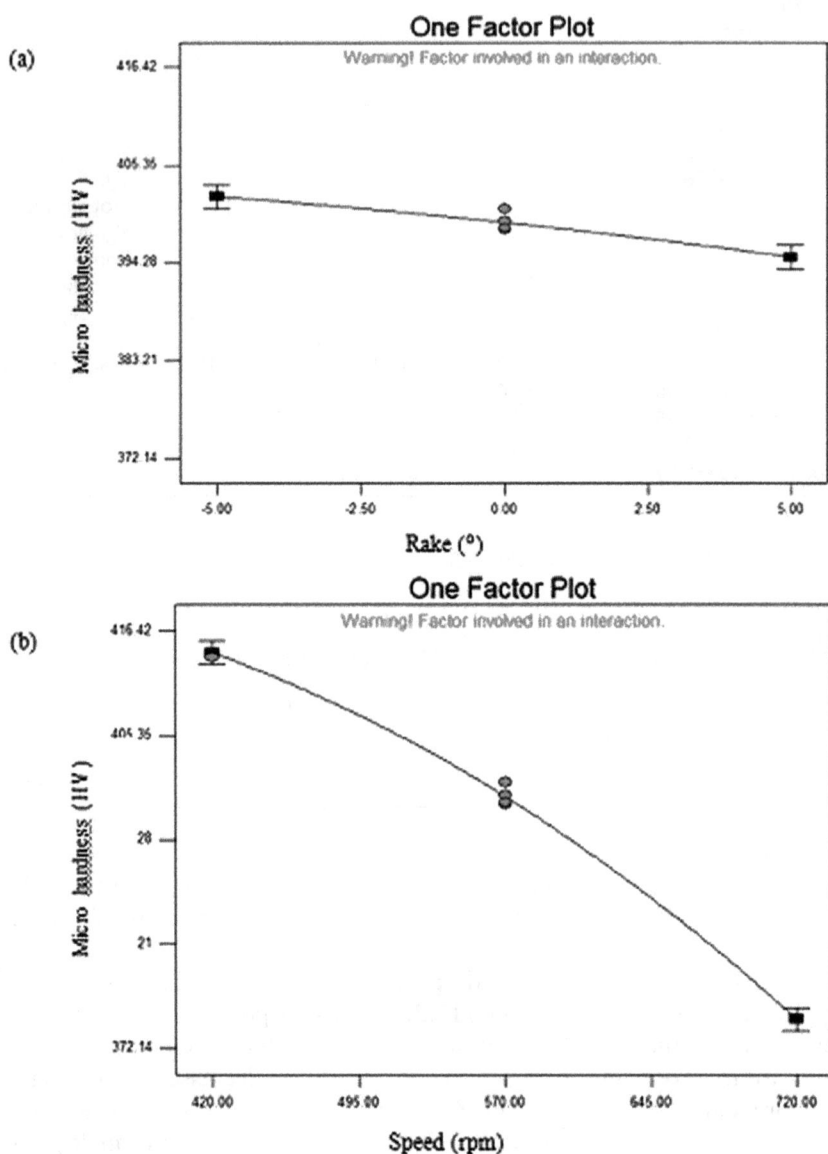

Figure 11.2 Variation of (a) microhardness vs. rake and (b) microhardness vs. speed.

have studied the effect of cutting speed on hardness and deformation temperature, and a similar trend was observed in their study. (See Figure 11.2.)

11.4.2 Analysis for surface roughness:

A Mitutoyo SJ-400 tester was used for the measurement of surface roughness. The tester uses a stylus method of measurement, has a profiler solution of 12nm, and can measure surface roughness up to 100μm. In this experiment, a tracing length of 4.8 mm was used.

The equation developed for response surface roughness is

$$\text{Surface roughness} = 3.073 + 0.0228 \times A - 0.0021 \times B - 3.625 \times C - 0.05 \times A \times A + 0.0000007 \times B \times B + 17 \times C \times C - 0.0003 \times A \times B + 0.29 \times A \times C - 0.0017 \times B \times C \qquad (11.2)$$

See also Tables 11.8 and 11.9.

The dependency of surface roughness on cutting speed is shown in Figure 11.3. The surface roughness of the samples decreases with an increase in cutting speed and rake angle, and surface roughness initially increases and then decreases with an increase in feed. The reason is that, at low cutting speeds, the matrix material is deformed to a lesser extent, leading to discontinuous chip formation with cracks and higher surface roughness. With increasing rake angle, the cutting forces and power consumption are reduced, causing less heat generation and tool wear, which further promotes continuous chip flow with reduced surface roughness.

11.4.3 Result for SEM

It is evident from Figure 11.4 that there is formation of sawtooth chip. If we compare the saw-tooth chip at different cutting speeds, then it is found that there is a transition of sawtooth from aperiodic at low speed to periodic at high speed.

11.5 CONCLUSIONS

On the basis of work conducted on Ti6Al4V titanium alloy chips, mechanical and microstructural properties were studied and the following conclusions were drawn:

1. A valid model was obtained for responses such as surface roughness and micro hardness.
2. With an increase in rake angle shear strain decreases and shear strain increases with an increase in the chip compression ratio.

Table 11.8 ANOVA table for surface roughness

Source	Sum of squares	DF	Mean square	F-value	Prob>F	Contribution (%)	Status
Model	6.89	9	0.77	87.09	<0.0001		Significant
A	0.049	1	0.049	5.61	0.0497	6.62	Significant
B	0.090	1	0.090	10.20	0.0152	1.21	Significant
C	0.010	1	0.010	1.14	0.3204	0.13	Not significant
A^2	6.58	1	6.58	747.91	<0.0001	88.99	Significant
B^2	0.0002632	1	0.000263	0.00299	0.9579	0.002	Not significant
C^2	0.007605	1	0.007605	0.86	0.3834	0.09	Not significant
AB	0.034	1	0.034	3.89	0.0892	0.473	Not significant
AC	0.021	1	0.021	2.39	0.1660	0.28	Not significant
BC	0.0001	1	0.0001	0.011	0.9181	0.001	Not significant
Residual	0.062	7	0.008796				
Lack of fit	0.046	3	0.015	3.93	0.1096		Not significant
Pure error	0.16	4	0.0039			2.16	
Cor total	6.96	10					

Significant factor: A, B, A^2

Insignificant factor: C, B^2, AB, AC, BC

Table 11.9 Different ANOVA parameters

R-squared	0.9911
Adj R-squared	0.9798
Pred R-squared	0.8908
Adeq precision	22.260

Notes:

The "Pred R-squared value" of 0.8908 is reasonable agreement with the "Adj R-squared value" of 0.9798.

"Adequate precision "value measure signal-to-noise ratio. It is desirable to have the ratio greater than 4. Ratio of 22.260 indicates adequate signal. Hence model can be used.

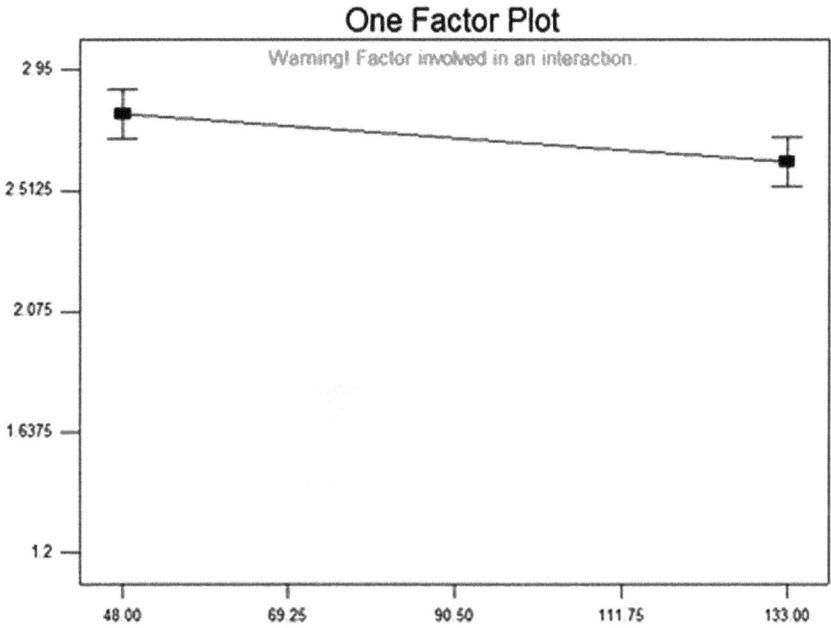

Figure 11.3 Variation of surface roughness vs. speed.

3. Microhardness of the fabricated chips was higher than that of ingot Ti6Al4V titanium alloy.
4. The increase in deformation level in chips is due to the increase in strain, resulting in decreased crystalline size.
5. The crystalline size of the fabricated chips is less than that of the Ti alloy, hence enhancing the mechanical properties.
6. The formation of sawtooth in the primary deformation zone is due to thermoplastic instability.

(a)

(b)

Aperiodic sawtooth

Periodic sawtooth

Figure 11.4 SEM image of sample showing (a) aperiodic sawtooth and (b) periodic sawtooth.

REFERENCES

1. H. Gleiter. (2000) Nanostructured materials: basic concepts and microstructure. *Acta Mater*, 48:1–29.
2. David D. Gill, Pin Yang, Aaron C. Hall, Tracy J. Vogler, Timothy J. Roemer, D. Anthony Fredenburg, Christopher J. Saldana. (2008) Creating Bulk Nanocrystalline Metal, SAND2008- 6547:1–92.
3. R.M. German. (1994) Source Powder Metallurgy Science. *Metal Powder Industries Federation*. ISBN 1-878954-42-3:1–472.
4. S.H. Whang. (2011) Nanostructured metals and alloys. Woodhead Publishing. ISBN: 978-1-84569-6702:1–840.
5. Nieman, G.W., J.R.Weertman, and R.W.Siegel. (1989) Microhardness of nanocrystalline palladium and copper produced by inert-gas condensation. *Scripta Metallurgica*, 23:2013–2018.
6. H.J. Fecht, E. Hellstern, Z. Fu, W.L. Johnson. (1990) Nano crystalline metals prepared by high-energy ball milling. *Material Transcations A*, 21:2333.
7. M. Ravi Shankar, Balkrishna C. Rao, Seongeyl Lee, Srinivasan Chandrasekar, Alexander H. King, W. Dale Compton. (2006) Severe plastic deformation (SPD) of titanium at near-ambient temperature. *ActaMaterialia*, 54:3691–3700.
8. M. Ravi Shankar, Srinivasan Chandrasekar, Alexander H. King, W. Dale Compton. (2005) Microstructure and stability of nanocrystalline aluminum 6061 created by large strain machining. *ActaMaterialia*, 53:4781–4793.
9. J.K. Jim, H.G. Jeong, S. I.Hong, Y.S. Kim and W.J. Kim.
10. S. Shekhar, S. Abolghasem, S. Basu, J. Caiand, M.R. Shankar. (2012) Effect of Severe Plastic Deformation in Machining Elucidated via

Rate-Strain-Microstructure Mappings. *Journal of Manufacturing Science and Engineering*, 134:031008-1.

11. S. Shekhar, S. Abolghashem, S. Basu, J.Cai, and M. Ravi Shankar. (2012) Interactive Effects of Strain, Strain-Rate and Temperatures on Microstructure Evolution in High Rate Severe Plastic Deformation. *Materials Science Forum*, 702-703:139–142.

12. C. Huang, T.G. Murthy, M.R. Shankar, R.M. Saoubi and S. Chandrasekar. (2008) Temperature rise in severe plastic deformation of titanium at small strain-rates. *ScriptaMaterialia*, 58:663–666.

13. S. Swaminathan, T.L. Brown, S. Chandrasekar, T.R. McNelley, and W.D. Compton. (2007) Severe plastic deformation of copper by machining: Microstructure refinement and nanostructure evolution with strain. *ScriptaMaterialia*, 56: 1047–1050.

14. R. Calistes, S. Swaminathan, T.G. Murthy, C. Huang, C. Saldana, M.R. Shankarand S. Chandrasekar. (2009) Controlling gradation of surface strains and nanostructuring by large-strain machining. *ScriptaMaterialia*, 60:17–20.

15. Andrew Kustas, Kevin Chaput, Srinivasan Chandrasekar, Kevin Trumble. (2014) Quality of strips produced by extrusion machining directly from Cast 5052 Aluminum. *Material Science and Technology*, 2014:79–86.

16. W.J. Deng, Q. Li, B. Li, Z.C. Xie, Y.T. He, Y. Tang and W. Xia. (2014) Thermal stability of ultrafine grained aluminium alloy prepared by large strain extrusion machining. *Materials Science and Technology*, 30(7):850–859.

17. Sharma, V.K., Kumar, V., Singh Joshi, R. (2020) Quantitative analysis of microstructure refinement in ultrafine-grained strips of Al6063 fabricated using large strain extrusion machining. *Machining Science and Technology*, 2020 Jan 2;24(1):42–64.

18. Dinakar Sagapuram, Andrew B. Kustas, W. Dale Compton, Kevin P. Tumble, Srinivasan Chandrasekar. (2015) Direct single-stage processing of light weight alloys into sheet by hybrid cutting extrusion. *Journal of Manufacturing Science and Engineering*, 137:1–10.

19. Travis L. Brown, Christopher Saldana, Tejas G. Murthy, James B. Mann, Yang Guo, Larry F. Allard, Alexander H. King, W. Dale Compton, Kevin P. Trumble, Srinivasan Chandrasekar. (2009) A study of the interactive effects of strain, strain rate and temperature in severe plastic deformation of copper. *ActaMaterialia*, 57:5491–5500.

Chapter 12

Role of ABD matrix model in the 3D study of laminate structures

Mansingh Yadav and Vipin Kumar Sharma

12.1 INTRODUCTION

Composite material has the capability of replacing conventional materials in military, aerospace, and other structural applications in which the weight of the structure is a major constraint. Laminated composite materials are currently employed in everything from domestic furniture to aerospace-grade structural parts due to their great strength and low density. Laminates are stiff and lightweight, with material qualities that may be tailored and producing constructions that are better than those produced of homogeneous materials [1–8].

The thickness of laminated composites depends on the number of laminae stacking over one another, and the mechanical properties of the laminate would be affected by the orientation of the laminas. The characteristics of the material combination, which are the attributes of the constituent layers, are unique. These features can be calculated and adjusted using traditional lamination concepts by changing the direction of different layers and the number of layers [9]. Most mechanical properties are dependent on the basic and simple materials defined as the ABD matrix in classical stacking theory. The ABD matrix records the relationship between the applied force/torque and the subsequent stretch, shear, bending, and twisting of the laminate [10].

However, the presence of multiple layers creates a complex shape and material variation in the laminate structure, and treating each layer separately makes structural analysis of the laminate very expensive. The standard

DOI: 10.1201/9781003360001-12

method assumes that the layers are fixed and that changes in the stress–strain field at the interface of the layers are ignored [11–14]. These assumptions allow you to simulate the global behavior of a laminate as a slab or shell behavior. Interlayer stresses and strains can be important in boundary and discontinuous regions [15] where a full 3D and/or layered approach must be used. It provides reasonably actual stress and strain calculation for the outer region. [16].

In composite panel design, the use of stacked variables is a two-step approach. The first step is to use the lamination parameters as optimization variables to create an optimization solution for panel stiffness, buckling strength, natural frequency, or other properties. Then use the obtained parameters in the second stage to create a price tacking sequence with properties very similar to the optimized plate properties. Stacking stiffness can be described as a linear combination of 12 stacking parameters [17], and the range of feasible stacking parameters is convex [18]. Instead of using the layer angle itself, the laminate design process may use the stiffness term contained in the laminate's ABD matrix. Realizability is also difficult to maintain with this "direct stiffness" modeling approach, and understanding the stacking sequence to obtain the appropriate ABD values is the final step.

Spacecraft structures such as deployable booms, reflector antennas, morph wings, and hinges are often thin-walled or shell-shaped for aerospace applications, enabling the widespread usage of single-layer woven reinforced composites (SWRC) [19–20]. SWRC is usually composed of weaving threads that are much thinner than the total thickness of the material, resulting in significantly uneven thickness. In addition, the mechanical properties of fiber-reinforced plastics and polymer resins are very different. The various tensile, bending, shearing, and torsional mechanical properties of composites are due to the nonuniformity and high behavior of their ultrastructure. As a result, describing complex constitutive properties seems to be more difficult compared to their isotropic mental counterparts [21–22]. The calculation should take into account the various properties of composite materials [23–24].The ABD stiffness factor connects the force and moment representation applied to apply macrostrains and curvatures so that many points can explain the mechanical properties of the shell model. By providing associated elements to A, B, and D matrices, the anisotropic shell configuration parameter of the composite material can be determined.

In this chapter, laminator is used to calculate the ABD matrix, which is a Windows 7/8/10 engineering program that analyses the laminated composite plate by conventional laminated plate theory. The input includes the plies' material properties, ply fiber orientation, material strength, stacking sequence, applied mechanical stresses or strains on the material, temperature, and moisture loading. The output includes ply stiffness and compliance matrices, laminate ABD matrices, ply stresses, laminate loads, and midplane strains, material axes, global strains, and load factors for ply failure using the Maximum Stress, Maximum Strain, Tsai-Hill, Hoffman, and Tsai-Wu failure theories. There is also a micromechanics calculator for calculating lamina properties based on fiber and matrix properties. For batch processing, the software also can also be run from the command line [25].

12.2 MATERIALS AND METHODS

12.2.1 Macromechanics of composite material

Classical Laminating Theory (CLT) is used to get the performance of composite laminates and determine the relationship between the external loading (moments M and forces N) with displacement (strains e and curvatures k). Several assumptions are made in the calculation of CLT. First, each layer of fibers should be parallel and continuous, even layers can have different thicknesses and materials. The second thickness of laminate should be much smaller than width and length dimensions, and the thickness's stress and strain are ignored. Next, it is assumed that bonding between the fibers and matrix and between the layers should be perfect [26–27]. The calculation is started in the lamination plate theory by first getting the stiffness in the longitudinal and transverse directions for each layer. If the properties are not adjusted in the global lamination orientation, then rotate and adjust the properties by using transformation and Reuter matrixes. Next, determine the top and bottom layer positions concerning the midplane of the laminate. After determining the transformed reduced stiffness matrix, it is used with simple formulae for the ABD matrix calculation in the laminate. Collectively, this ABD stiffness matrix is used to get the relationship between the external loading (moments M and forces N) and displacement (strains e and curvatures k) that is shown in equation (12.7). (See Tables 12.1 and 12.2.)

The ABD matrix is used to describe the homogenized properties of the material in the classical lamination theory of macromechanics because in the microstructural heterogeneity composite analyses, individual properties of the constituents are very complex so macroanalysis is used to get the

Table 12.1 Prepregs specifications (size, grades, GSM, and thickness)

Size (250* 300) mm²	Spectra grade	GSM (gm²)	Thickness/layer (mm)
0.075	3124	220	0.282
0.075	5143	165	0.214
0.075	6472	115	0.126

Table 12.2 Panel specifications with the combination of different grades

Spectra grade	Thickness (mm)	No. of layers	Areal density (AD) kg/m²
3124	3.5	13	2.86
5143	1.5	7	1.15
6472	1	8	0.92
Total	6	28	4.93

Table 12.3 Material properties of UHMWPE (ultrahigh molecular weight polyethylene for calculation (31-33)

Grade	E_1 (GPa)	E_2 (GPa)	G_{12} (GPa)	μ_{12}
3124	24	1.4	0.7	0.1
5143	30	1.75	0.875	0.1
6472	36	2.1	1.05	0.1

average behavior properties of the constituents. The formulae used in the classical lamination theory are:

$$
\begin{bmatrix} \sigma_x \\ \sigma_y \\ \tau_{xy} \end{bmatrix} = [T]^{-1} \begin{bmatrix} \sigma_x \\ \sigma_y \\ \tau_{xy} \end{bmatrix} = [T]^{-1}[Q][R][T][R]^{-1} \begin{bmatrix} \varepsilon_x \\ \varepsilon_y \\ \gamma_{xy} \end{bmatrix} \tag{12.1}
$$

$$
\begin{bmatrix} \sigma_x \\ \sigma_y \\ \tau_{xy} \end{bmatrix} = [\bar{Q}] \begin{bmatrix} \varepsilon_x \\ \varepsilon_y \\ \gamma_{xy} \end{bmatrix} = \begin{bmatrix} \bar{Q}_{11} & \bar{Q}_{12} & \bar{Q}_{16} \\ \bar{Q}_{12} & \bar{Q}_{22} & \bar{Q}_{26} \\ \bar{Q}_{16} & \bar{Q}_{26} & \bar{Q}_{66} \end{bmatrix} \begin{bmatrix} \varepsilon_x \\ \varepsilon_y \\ \gamma_{xy} \end{bmatrix}
$$

$$
\bar{Q}_{11} = Q_{11} \cos^4 \theta + 2(Q_{12} + 2Q_{66}) \sin^2 \theta \cos^2 \theta + Q_{22} \sin^4 \theta
$$

$$
\bar{Q}_{12} = (Q_{11} + Q_{22} - 4Q_{66}) \sin^2 \theta + Q_{12} (\sin^4 \theta + \cos^4 \theta)
$$

$$
\bar{Q}_{22} = Q_{11} \sin^4 \theta + 2(Q_{12} + 2Q_{66}) \sin^2 \theta \cos^2 \theta + Q_{22} \cos^4 \theta
$$

$$
\bar{Q}_{16} = (Q_{11} - Q_{12} - 2Q_{66}) \sin \theta \cos^3 \theta + (Q_{12} - Q_{22} + 2Q_{66}) \sin^3 \theta \cos \theta
$$

$$
\bar{Q}_{26} = (Q_{11} - Q_{12} - 2Q_{66}) \sin^3 \theta \cos \theta + (Q_{12} - Q_{22} + 2Q_{66}) \sin \theta \cos^3 \theta
$$

$$\bar{Q}_{66} = (Q_{11} + Q_{12} - 2Q_{12} - 2Q_{66}) \sin^2\theta\cos^2\theta + Q_{66}(\sin^4\theta\cos^4\theta) \quad (12.2)$$

$$[T] = \begin{bmatrix} \cos^2\theta\sin^2\theta & 2\sin\theta\cos\theta \\ \sin^2\theta\cos^2\theta & -2\sin\theta\sin\theta \\ -\sin\theta\cos\theta\sin\theta\cos\theta\cos^2\theta - \sin^2\theta \end{bmatrix} \quad (12.3)$$

$$[R] = \begin{bmatrix} 1 & 0 & 0 \\ 0 & 1 & 0 \\ 0 & 0 & 2 \end{bmatrix} \quad (12.4)$$

Here equation (12.3) shows the transformation matrix that is used to rotate the stress direction, and equation (12.4) shows the Reuter matrix that is used to get shear strain in classical lamination because it is twice the tensorial shear strain.

$$Q_{11} = \frac{E_1}{1 - v_{12}v_{21}}$$

$$Q_{22} = \frac{E_2}{1 - v_{12}v_{21}}$$

$$Q_{66} = G_{12}$$

$$Q_{12} = \frac{v_{12}E_2}{1 - v_{12}v_{21}} = \frac{v_{21}E_1}{1 - v_{12}v_{21}}$$

$$Q_{66} = G_{12} \quad (12.5)$$

Laminate stiffeness matrix [A, B, D] sequences:

$$A_{ij} = \sum_{k=1}^{n} \left[\bar{Q}_{ij} \right]_k (h_k - h_{k-1})$$

$$B_{ij} = \frac{1}{2} \sum_{k-1}^{n} \left[\bar{Q}_{ij} \right]_k (h^2_k - h^2_{k-1})$$

$$D_{ij} = \frac{1}{3} \sum_{k=1}^{n} \left[\bar{Q}_{ij} \right]_k (h^3_k - h^3_{k-1}) \quad (12.6)$$

$$
\begin{Bmatrix} N_x \\ N_y \\ N_{xy} \\ -- \\ M_x \\ M_y \\ M_{xy} \end{Bmatrix} =
\left(\begin{array}{ccc|ccc}
A_{11} & A_{12} & A_{16} & B_{11} & B_{12} & B_{16} \\
A_{21} & A_{22} & A_{26} & B_{21} & B_{22} & B_{26} \\
A_{61} & A_{62} & A_{66} & B_{61} & B_{62} & B_{66} \\
\hline
B_{11} & B_{12} & B_{16} & D_{11} & D_{12} & D_{16} \\
B_{21} & B_{22} & B_{62} & D_{21} & D_{22} & D_{26} \\
B_{61} & B_{62} & B_{66} & D_{61} & D_{62} & D_{66}
\end{array} \right)
\begin{Bmatrix} \varepsilon_x \\ \varepsilon_y \\ \gamma_{xy} \\ -- \\ \kappa_x \\ \kappa_y \\ \kappa_{xy} \end{Bmatrix} \quad \text{Eq. (7)}
$$

Applied loads & movements   Stiffness matrix   Strains

N, M, ε, and K are loads, movements, strain, and curvature, respectively, where A_{ij} is the extensional and shear stiffnesses. B_{ij} is the extension-bending coupling stiffnesses, and its nonzero value show the coupling between twisting/bending curvatures and shear/extension load and for symmetric laminates its value would be zero. For dimensional stability, the laminate is designed symmetrically. D_{ij} shows the torsional and bending stiffnesses. Equation (12.8) shows the effect of individual elements in the matrix.

$$
\begin{Bmatrix} N_x \\ N_y \\ N_{xy} \\ M_x \\ M_y \\ M_{xy} \end{Bmatrix} =
$$

A_{11} ISO	A_{12} ISO	A_{16} SE	B_{11} BE	B_{12} BE	B_{16} ET	ε_x
A_{22} ISO	A_{22} ISO	A_{26} SE	B_{21} BE	B_{22} BE	B_{26} ET	ε_y
A_{61} SE	A_{62} SE	A_{66} ISO	B_{61} BS	B_{62} BS	B_{66} ST	γ_{xy}
B_{11} BE	B_{12} BE	B_{16} BS	D_{11} ISO	D_{12} ISO	D_{16} BT	κ_x
B_{21} BE	B_{22} BE	B_{26} BS	D_{21} ISO	D_{22} ISO	D_{26} BT	κ_y
B_{61} ET	B_{62} ET	B_{66} ST	D_{61} BT	D_{62} BT	D_{66} ISO	κ_{xy}

(12.8)

where

ISO = isotropic material response, ET = twist (torsion)-extension coupling,
SE = shear-extension coupling, ST = shear-twist (torsion) coupling,
BE = bend-extension coupling, BT = bend-twist (torsion) coupling,
BS = bend-shear coupling,

Matrix A is known as an extensional stiffness matrix that is related to normal strains and stresses. Matrix B has used the combination of coupled stiffnesses including bend-extension, twist-extension, shear-extension, and bend-shear. Matrix D is known as the bending stiffness matrix, which relates the curvatures with bending moments [28–30]. See Figure 12.1.

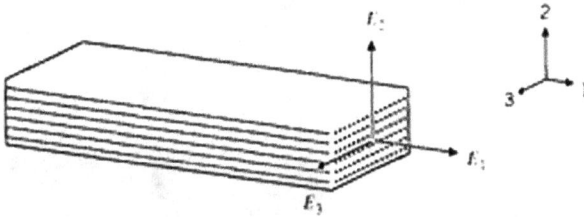

Figure 12.1 Orthotropic material [25].

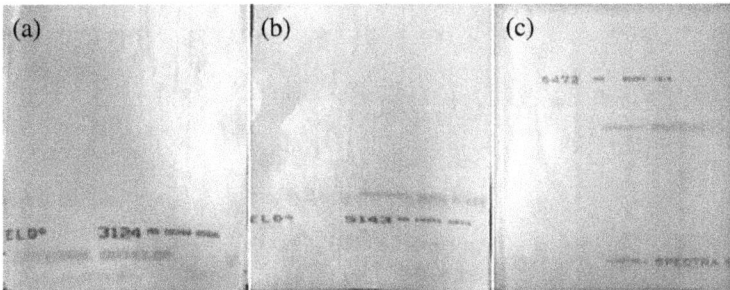

Figure 12.2 Different grades of UHMWPE: (a) Spectra® 3124, (b) Spectra® 5143, (c) Spectra® 6472.

12.2.2 Material

UHMWPE material contains a long chain that provides high tensile strength and tenacity. The outstanding mechanical properties of UHMWPE make it suitable for ballistic material. Its molecular mass is high between 3.5 and 7.5 million amu. The structure of UHMWPE is where n is more than 100,000 in UHMWPE. Tenacity, tensile strength, and impact strength are some important properties for the ballistic application. (See Figure 12.2. and 12.3.)

All ply stacks were consolidated using a vacuum-assisted compression molding machine; these were the next step in the fabrication of ballistic material with specified consolidation pressure and temperature. It was a multistep process that occurs in a single stroke: the hydraulic pump is used to apply pressure, the heating coil is used to heat the ram and bed, vacuum pumps generate vacuum in the chamber, and the water pump is used for cooling the setup. Two main controlling parameters in a vacuum compression molding machine are pressure and temperature with which we can control the quality of the laminate panels. Here it is necessary to get the required quality product by proper sequencing of the plies using Laminator software.

Figure 12.3 SEM images: (a) Spectra® 3124, (b) Spectra® 5143 fibers.

12.3 RESULTS AND DISCUSSION

A combination of different grades is formed in unsymmetrical laminates that contain nonzero coupling matrix B, which indicates that the composite laminate exhibits bending and twisting, even if the external moment on it is perfectly zero. Table 2.4 shows the two elements in the matrix B are B_{12} (9372 MPa mm²) and B_{21} (9372 MPa mm²), which indicates coupling effects due to nonuniform lateral effects. (See Tables 12.4 and 12.5.)

See also Table 12.6.

Here two elements values are B_{12} (–9031 MPa mm²) and B_{21} (–9031 MPa mm²) in the B matrix, which indicates coupling effects due to nonuniform lateral effects. The effect of both sequences are the same, but the matrix element B_{12} and B_{21} are positive in the Spectra® 3124, 5143, and 6472 sequence and negative in the Spectra® 6472, 5143, and 3124, so the coupling effect would be opposite each other in direction.

Table 12.4 ABD matrix for Spectra® 3124, 5143, and 6472 laminate sequence

A matrix			B matrix		
1.932el I	1.932el0	−1.329e(−4)	9.372el0	9.370e9	−8.094e(−5)
1.932el0	1.932el I	1.329e(−4)	9.372e9	9.372el0	8.094e(−5)
−1.329e(−4)	1.329e(−4)	4.935e9	−8.094e(−5)	8.094e(−5)	1.259e9

D matrix		
6.970el I	6.970el0	−4.806e(−4)
6.970el0	6.970el I	4.806e(−4)
−4.806e(−4)	4.806e(−4)	1.660el0

A inverse matrix			B inverse matrix		
5.59 le(−12)	−5.59 le(−13)	1.535e(−25)	−7.517e(−13)	7.517e(−14)	−5.599e(−27)
−5.59 le(−13)	5.59 le(−12)	−1.535e(−25)	7.517e(−14)	−7.517e(−13)	5.599e(−27)
1.535e(−25)	−1.535e(−25)	2.066e(−10)	−5.983e(−27)	5.983e(−27)	−1.566e(−11)

D inverse matrix		
1.550e(−12)	−1.550(−13)	4.577e(−26)
−1.550e(−13)	1.550e(−12)	−4.577e(−26)
4.577e(−26)	−4.577e(−26)	6.139e(−11)

Table 12.5 Effect of laminate sequence Spectra® 3124, 5143, and 6472

Matrix element	Approximate	Effect of laminate sequence
$A_{16} = A_{26}$	0	Balance laminate
B_{ij}	Not zero	Unsymmetrical laminate
$A_{16} = A_{26}$ and $D_{16} = D_{26}$	0	Antisymmetrical laminate

ABD matrix Spectra® 6472, 5143 and 3124 laminate sequence

A matrix			B matrix		
1.894el I	1.894el0	−1.269e(−4)	−9.03 lel0	−9.03 le9	5.682e(−5)
1.894el0	1.894el I	1.269e(−4)	−9.03 le9	−9.013el0	−5.682e(−5)
−1.269e(−4)	1.269e(−4)	4.826e9	−5.682e(−5)	−5.682e(−5)	−1.21 le9

D matrix		
6.475el I	6.475el0	−4.228e(−4)
6.475el0	6.475el I	4.228e(−4)
−4.228e(−4)	4.228e(−4)	1.540el0

(continued)

Table 12.5 Cont.

Inverse A Matrix			Inverse B Matrix		
$5.710e(-12)$	$-5.710e(-13)$	$1.583e(-25)$	$7.964e(-13)$	$-7.964e(-14)$	$1.333e(-26)$
$-5.710e(-13)$	$5.710e(-12)$	$-1.83e(-25)$	$-7.964e(-14)$	$7.964e(-13)$	$-1.333e(-26)$
$1.583e(-25)$	$-1.583e(-25)$	$2.113e(-10)$	$1.353e(-26)$	$-1.353e(-26)$	$1.662e(-11)$

Inverse D Matrix		
$1.671e(-12)$	$-1.671e(-13)$	$4.829e(-26)$
$-1.671e(-13)$	$1.671e(-12)$	$-4.829e(-26)$
$4.829e(-26)$	$-4.829e(-26)$	$6.623e(-11)$

Table 12.6 Effect of laminate sequence Spectra® 6472, 5143, and 3124

Matrix element	Approximate	Effect of laminate sequence
$A_{16} = A_{26}$	0	Balance laminate
B_{ij}	Not zero	Unsymmetrical laminate
$A_{16} = A_{26}$ and $D_{16} = D_{26}$	0	Antisymmetric laminate

12.4 CONCLUSIONS

The Classical Laminating Theory (CLT) is used to evaluate the perform-ance of composite laminate and determine the relationship between external loads (moments M and forces N) and displacements (strain e and deflection k). By means of the ABD matrix, it is possible to get the effect of the stacking order on the mechanical properties. The thickness of laminated composites depends on the number of lamina stacking over one another, and the mech-anical properties of the laminate would be affected by the orientation of the laminas. The ABD matrix is essential to get the effect on mechanical proper-ties by knowing the stacking sequence. The ABD matrix value is calculated in this chapter using a laminator as a function of the number of layers and its orientation angle. The relationship between in-plane force and in-plane strain is represented by elongation stiffness (A). In-plane force and in-plane deflection relationship is represented by coupling stiffness (B). The letter D stands for flexural rigidity and links the generated moment to the curvature of the midplane. The presence of nonzero elements in the coupling matrix B indicates that the composite laminate will exhibit twisting and bending, even if the externally applied moments on it are zero. This chapter kept the layers in 0/90°alignment with the number of layers and got the A, B, D matrix values and their behavior on the laminate. The effects of both sequences are the same, but the matrix element B_{12} and B_{21} are positive in the Spectra® 3124, 5143, and 6472 sequence and negative in the Spectra® 6472, 5143, and 3124, so the coupling effects would be opposites in direction.

REFERENCES

[1] Barbosa, A., Upadhyaya, P., & Iype, E. (2020). Neural network for mechanical property estimation of multilayered laminate composite. Materials Today: Proceedings, 28, 982–985.

[2] F.C. Campbell Jr., Manufacturing processes for advanced composites, Access Online via Elsevier, 2003.

[3] A. Handbook, vol. 21, Composites 380.

[4] I.M. Daniel, O. Ishai, I.M. Daniel, I. Daniel, Engineering Mechanics of Composite Materials, vol. 3, Oxford University Press, New York, 1994.

[5] Sharma, V.K., Kumar, V., Joshi, R.S. Effect of RE addition on wear behavior of an Al-6061 based hybrid composite. Wear. 2019 Apr 30;426:961–74.

[6] Sharma, V.K., Kumar, V., Joshi, R.S. Investigation of rare earth particulate on tribological and mechanical properties of Al-6061 alloy composites for aerospace application. Journal of Materials Research and Technology. 2019 Jul 1;8(4):3504–16.

[7] Sharma, V.K., Kumar, V., Joshi, R.S. Experimental investigation on effect of RE oxides addition on tribological and mechanical properties of Al-6063 based hybrid composites. Materials Research Express. 2019 Jun 5;6(8):0865d7.

[8] Sharma, V.K., Aggarwal, D., Vinod,, K., Joshi R.S. Influence of rare earth particulate on the mechanical & tribological properties of Al-6063/SiC hybrid composites. Particulate Science and Technology. 2021 Nov 17;39(8):928–43.

[9] C.W. Bert, Class. Lamin. Theory (1989) 11–16.

[10] C.T. Herakovich, Composite materials: lamination theory, in: Luigi Nicolais, Assunta Borzacchiello (Eds.), Wiley Encycl. of Composites, Second Ed., 2012, pp. 1–5.

[11] A. Handbook, vol. 21, Composites 380.

[12] I.M. Daniel, O. Ishai, I.M. Daniel, I. Daniel, Engineering Mechanics of Composite Materials, vol. 3, Oxford University Press, New York, 1994.

[13] J.N. Reddy, Mechanics of Laminated Composite Plates and Shells: Theory and Analysis/JN Reddy, CRC Press, Boca Raton, 2004.

[14] F.L. Matthews, Finite Element Modelling of Composite Materials and Structures, CRC Press, Boca Raton, 2000.

[15] K.J. Saeger, P.A. Lagace, D.J. Shim, Interlaminar stresses due to in-plane gradient stress fields, J.Compos.Mater. 36(2)(2002) 211–227.

[16] J. Reddy, An evaluation of equivalent-single-layer and layerwise theories of composite laminates, Compos.Struct. 25(1)(1993) 21–35.

[17] Tsai, S., Hahn, H. Introduction to composite materials. Technomic Publishing Company, Inc.; 1980.

[18] Grenestedt, J., Gudmundson, P. Layup optimization of composite material structures. Elsevier Science Amsterdam 1993:311–36.

[19] Santo, L., Quadrini, F., Ganga, P.L., Zolesi, V. Mission BION-M1: Results of RIBES/FOAM2 experiment on shape memory polymer foams and composites. Aerosp SciTechnol 2015;40:109–14.

[20] Liu, Y., Du, H., Liu, L., Leng, J. Shape memory polymers and their composites in aerospace applications: a review. Smart Mater Struct 2014;23: 023001.

[21] Gao, J., Chen, W., Fan, P., Zhao, B., Hu, J., Zhang, D., et al. Experimental determination of mechanical properties of a single-ply broken twill 1/3 weave reinforced shape memory polymer composite. Polym Test 2018;69:100–6.

[22] Gao, J., Chen, W., Yu, B., Fan, P.., Zhao, B, Hu, J., et al. Effect of temperature on the mechanical behaviours of a single-ply weave-reinforced shape memory polymer composite. Compos Part B Eng 2019;159: 336–45.

[23] Daniel, I.M., Ishai, O. Engineering Mechanics of Composite Materials. New York: Oxford University Press; 2006.

[24] Kaw, A. Mechanics of Composite Materials. Boca Raton: CRC Press; 2006.

[25] Lindell, M. (2006). The laminator-classical analysis of composite laminates.

[26] https://www.pursuitcycles.com/2018/04/02/the-abds-of-composite-laminates/

[27] Bert, C.W. (1989). Classical lamination theory. In Manual on Experimental Methods for Mechanical Testing of Composites (pp. 11–16). Springer, Dordrecht.

[28] Kumar, G., & Shapiro, V. (2015). Efficient 3D analysis of laminate structures using ABD-equivalent material models. Finite Elements in Analysis and Design, 106, 41–55.

[29] Gao, J., Chen, W., Yu, B., Fan, P., Zhao, B., Hu, J., & Peng, F. (2019). A multiscale method for predicting ABD stiffness matrix of single-ply weave-reinforced composite. Composite Structures, 230, 111478.

[30] Jones, R.M. (1998). Mechanics of composite materials. CRC press.

[31] Dayyoub, T., Maksimkin, A.V., Kaloshkin, S., Kolesnikov, E., Chukov, D., Dyachkova, T.Y.P., & Gutnik, I. (2019). The structure and mechanical properties of the UHMWPE films modified by the mixture of graphene nanoplates with polyaniline. Polymers, 11(1), 23.

[32] Nguyen, L.H., Lässig, T.R., Ryan, S., Riedel, W., Mouritz, A.P., & Orifici, A.C. (2015). Numerical modelling of ultra-high molecular weight polyethene composite under impact loading. Procedia Engineering, 103, 436–443.

[33] Hazzard, M K., Trask, R.S., Heisserer, U., Van Der Kamp, M., & Hallett, S.R. (2018). Finite element modelling of Dyneema® composites: From quasi-static rates to ballistic impact. Composites Part A: Applied Science and Manufacturing, 115, 31–45.

Chapter 13

RSM-based surface and subsurface characterization analysis of bioimplant material

Anish Kumar, Vinod Kumar, Renu Sharma, and Jatinder Kumar

13.1 MACHINABILITY OF TITANIUM

In the medical and manufacturing industries, customization, or changing a product to meet a specific demand, has become a critical requirement. Many biomedical devices, notably prostheses and implants, restore damaged limbs or bones or restore precise joint function in a patient [1]. In contrast to stainless steel, titanium grade-2 is a good alternative for a range of devices being used in dental and orthopedic applications, as well as in the aerospace industry [2]. Such characteristics are essential when understanding how well the body's surroundings will respond to situations like sliding wear and cyclic stresses. When inappropriate chemicals are embedded in human bodies, the body might recognize them and attempt to separate them by placing them in fibrous tissue. On the other hand, titanium does not cause any adverse effects and is well accepted by body cells [3]. Titanium grade-2 is also

DOI: 10.1201/9781003360001-13

active in osseointegration (screw form) dental implant applications [4]. As a result, many biocompatible materials are made using traditional machining methods; however, these methods are ineffective for milling titanium [5–7]. Traditional machining produces difficult-to-machine materials, yet the material removal rate (MRR) is quite low, resulting in a higher tool wear rate. The necessity for reliable and efficient machining procedures for CP-Ti is critical. The wire electric discharge machining (WEDM) process seems to be a better selection because it can cut components with complex and intricate profiles. As a result, an effort has been made in this study to assess the machining characteristics of titanium grade-2 using the WEDMed process. The machining characteristics of this biocompatible material are less documented in the literature. Kuriakose and Shunmugam [8] applied a regression model to develop the relation between input and output variables. Multiobjective optimization of a nondominated sorting Genetic Algorithm was used. The sorting technique was employed to provide nondominated solutions. Sarkar et al. [9–11] established regression models to analyze the cutting speed, surface roughness, and dimensional deviation in WEDM of titanium aluminide alloy. They achieved the optimal process parameters by using a constrained optimization technique. Using the response surface methodology, Hseigh et al. [12] explored the properties of ternary shape memory alloys. Cydas et al. [13] developed a model to analyze the recast layer thickness (RLT) and average surface roughness by way of a function of process factors using an artificial fuzzy network (AFN) inference system. The performance of WEDM in aluminum-based composites was investigated by Kung and Chiang [14]. Huang et al. [15] investigated the surface characteristics of martensitic steel fine surface by wire electrical discharge machining (EDM). Liao and Yu [16] examined the specific discharge energies on Ti-6Al-4V, SS, and Inconel 718, reporting that the larger the discharge energy is, the smaller the spark gap will be. Using dimensional analysis, Poros and Zoboruski [17] derived a semiempirical relationship on thermal characteristics for tough titanium alloys. In the machining of Ti6Al4V, thermal appearances were the most significant factor. The properties of Ti-Ni-X ternary shape memory alloys were explored by Hseigh et al. [18]. With increasing pulse length, the surface roughness of machined Ti-Ni-X alloy increased, yielding in the outcomes. Garg et al. [19] used response surface methods to explore the WEDM parameters on Ti-6242 and used a Genetic Algorithm (GA) method to find the best configuration. Aspinwall et al. [20] used minimum damage generator technology to examine the surface characteristics of Ti-6Al-4V and Inconel-718 after a WEDM process. Kumar et al. (2013) [21–22] used RSM methodology to assess the WEDM parameters on pure titanium and exposed a relationship between machining rate and surface roughness. In the whole application of titanium in biomedical areas, machining of titanium grade-2 with high precision is a serious problem. A WEDM process uses electrical pulses to cut electrically conductive titanium grade-2 material. During this process, a discrete succession of electrical sparks was produced underneath the spark gap. The temperatures range between 10,000°C to 12,000°C. Erosion, melting, and vaporization are used to remove the material. For a brief moment, the molten material was washed away by deionized water [23]. Furthermore, repeated thermal heating and quenching lead to the formation of a recast layer. In comparison to the parent material, this layer has different microcracks, craters, and porosity. Microscopy is used to optically detect this recast layer. Several researchers have attempted to improve WEDM performance in alloys and composites by employing single-objective classical optimization techniques. The current study's uniqueness was centered on evaluating the surface and subsurface morphology of titanium grade-2 utilizing a multiobjective response surface methodology evolutionary technique to explore the effect of WEDMed parameters. The WEDMed surface was examined using scanning electron microscope (SEM), energy dispersive X-ray (EDX), and X-ray diffraction (XRD) spectroscopy techniques after machining.

13.2 MATERIAL AND METHODS

The experimental plan was carried out on a 4-axis computerized numerical control (CNC) type of WEDM in this research work, and steps 1–9 were followed as shown in Figure 13.1. Table 13.1 shows the list of factors and their levels. The levels for the factors were decided on the basis of a pilot study and past work. The chemical composition of the selected material is as follows: C has a concentration of 0.10%, N has a concentration of 0.03%, O has a concentration of 0.25%, H has a concentration of 0.015%, Fe has a concentration of 0.30%, and Ti has a concentration of 99.03%. Six input parameters, namely POT (pulse on time), POFT (pulse off time), PC (peak current), SGV, wire speed (WS), and wire tension (WT), were adapted based on a pilot and previous survey results. The experimental design matrix was developed through Design-Expert software 12.0 version using response-surface-methodology-based Box Behnken design (BBD) approach, and its outputs are shown in Table 13.2. Electrode (brass wire with 0.25 mmØ), t: 26 mm, and DWP (dielectric water pressure): 7 kg/cm² were the constant parameters. An SEM with an integrated EDX setup was used to study surface morphology. Acetone (CH_3)2CO was used to clean the 54 work samples. Further, EDX was used to assess material migration from a brass wire, deionized water, and a work sample. A scanning electron microscope was used to examine micrographs of samples at various magnification levels (ranging from 500× to 3000×). Before the RLT demonstration, the sample was cut off and cleaned sequentially with silicon carbide paper.

Table 13.1 Factors and their levels (coded and actual)

Factor	Name	Units	Lower range	Upper range	Coded low	Coded high	Mean	Std. dev.
A	POT	µs	0.7	1.1	1↔0.7	+1↔1.1	0.9	0.1346
B	POFT	µs	17	38	−1↔17	+1↔38	27.5	7.07
C	PC	A	120	200	1↔120	+1↔200	160	26.92
D	SGV	V	40	60	−1↔40	+1↔60	50	6.73
E	WS	m/min	4	10	−1↔4	+1↔10	7	2.02
F	WT	g	500	1400	1↔500	+1↔1400	950	302.82

Table 13.2 Design matrix for main experimentation

Run no.	POT (µs)	POFT (µs)	PC (A)	SGV (V)	WS (m/min)	WT (g)	Crack density (µm/µm^2)	RLT (µm)
1	1.1	28	200	50	7	500	0.025	58.63
2	0.9	38	160	50	4	500	0.005	46.47
3	0.7	28	160	60	4	950	0.008	31.01
4	0.9	17	120	50	10	950	0.009	49.35
5	0.9	28	120	60	7	500	0.007	44.48
6	1.1	28	160	40	4	950	0.014	49.35
7	0.9	38	160	50	10	1400	0.005	44.42
8	0.9	28	160	50	7	950	0.007	47.32
9	0.9	17	160	50	4	500	0.01	46.37
10	1.1	28	160	40	10	950	0.016	53.63
11	1.1	38	160	40	7	950	0.014	49.8
12	1.1	28	160	60	4	950	0.015	46.54
13	0.9	17	160	50	10	500	0.009	47.25
14	0.9	28	160	50	7	950	0.007	45.24
15	0.7	28	120	50	7	500	0.006	33.28
16	0.9	28	160	50	7	950	0.007	44.52
17	0.9	28	120	60	7	1400	0.007	44.45
18	0.7	38	160	40	7	950	0.007	32.75
19	0.9	38	120	50	10	950	0.006	44.21
20	0.9	28	200	40	7	1400	0.021	49.41
21	0.9	28	200	60	7	500	0.019	47.88
22	0.9	38	200	50	10	950	0.017	44.31
23	0.9	28	120	40	7	1400	0.009	48.32
24	0.7	28	120	50	7	1400	0.007	30.36
25	0.9	38	200	50	4	950	0.02	44.94
26	1.1	28	160	60	10	950	0.012	49.6
27	1.1	28	120	50	7	500	0.009	43.36
28	0.7	28	160	40	10	950	0.008	39.13
29	0.7	28	200	50	7	500	0.011	41.19
30	0.7	17	160	40	7	950	0.013	41.31
31	0.7	28	200	50	7	1400	0.012	31.21
32	0.9	28	160	50	7	950	0.007	45.16
33	0.9	17	200	50	4	950	0.017	53.94

Table 13.2 Cont.

Run no.	POT (μs)	POFT (μs)	PC (A)	SGV (V)	WS (m/min)	WT (g)	Crack density (μm/μm^2)	RLT (μm)
34	0.9	28	160	50	7	950	0.008	44.27
35	1.1	17	160	40	7	950	0.023	56.13
36	0.9	17	200	50	10	950	0.013	52.36
37	0.9	28	200	40	7	500	0.021	54.13
38	0.7	28	160	40	4	950	0.012	32.1
39	0.9	38	160	50	10	500	0.008	44.75
40	0.9	28	160	50	7	950	0.004	49.23
41	1.1	38	160	60	7	950	0.014	44.42
42	0.7	17	160	60	7	950	0.01	32.54
43	0.9	28	200	60	7	1400	0.02	46.21
44	0.9	17	120	50	4	950	0.015	42.64
45	0.7	28	160	60	10	950	0.008	33.72
46	1.1	28	120	50	7	1400	0.014	42.15
47	0.7	38	160	60	7	950	0.01	29.35
48	0.9	17	160	50	4	1400	0.014	44.36
49	0.9	28	120	40	7	500	0.013	42.39
50	1.1	17	160	60	7	950	0.019	46.56
51	0.9	38	120	50	4	950	0.004	43.98
52	1.1	28	200	50	7	1400	0.017	33.64
53	0.9	17	160	50	10	1400	0.012	46.28
54	0.9	38	160	50	4	1400	0.007	44.5

13.3 RESULTS AND DISCUSSION

13.3.1 WEDMed surface morphology analysis of subsurface of titanium grade-2

The present study is mainly concerned with the morphology of the surface and subsurface of WEDMed of titanium grade-2. The craters, pockmarks, debris, microcracks, and a recast layer were observed and significantly affected by the POT, POFT, and PC factors. During WEDMed, the discharge energy impinging on the workpiece melts the material, which is washed away by a dielectric. The surface morphology of a machined component was affected by the thermal action of the EDM process [24]. During the metal removal process, enormously rapid heating, melting, and vaporization were monitored, resulting in a significant change in the subsurface of CP-Ti. As shown in SEM micrographs (Figure 13.2), the WEDMed features include globules of debris, spherical particles, craters, pockmarks, and microcracks. The most important parameters responsible for the surface characteristics were found to be the POT (1.1 μs) and PC (200 A). The deep craters on the machined surface were produced by an increase in POT. As shown in Figure 13.2(f), molten material solidifies to create lumps of debris. As seen in Figure 13.3(a)–(c), the machined surface has hillocks and valleys

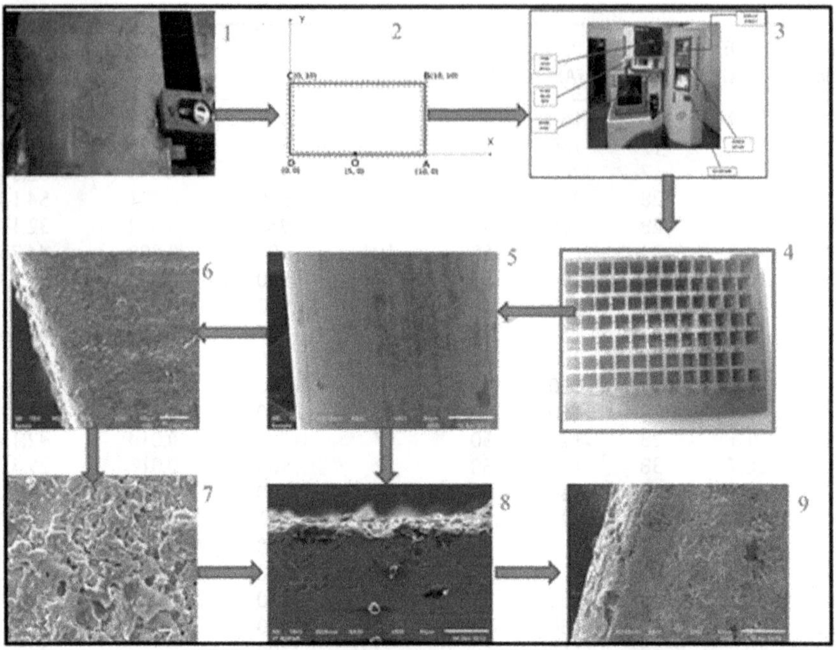

Figure 13.1 Steps 1–9: (1) holding fixture, (2) workpiece profile, (3) WEDM setup, (4) square punch workpiece profile, (5) fresh brass wire, (6) fresh work material, (7) work material after WEDMed, (8) RL at the cross section, (9) wear-out wire.

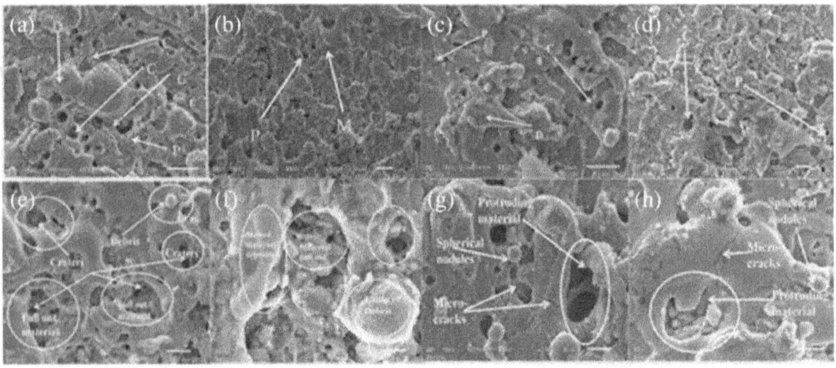

Figure 13.2 Micrographs observed with C → Craters, P → Pockmarks, D → Debris, → Matt surface, spherical nodule and protruding material at higher POT =1.1 µs, POFT= 17 µs, and PC = 200A, SR = 3.22 µm, 2.93 µm, 2.68 µm, 2.55 µm.

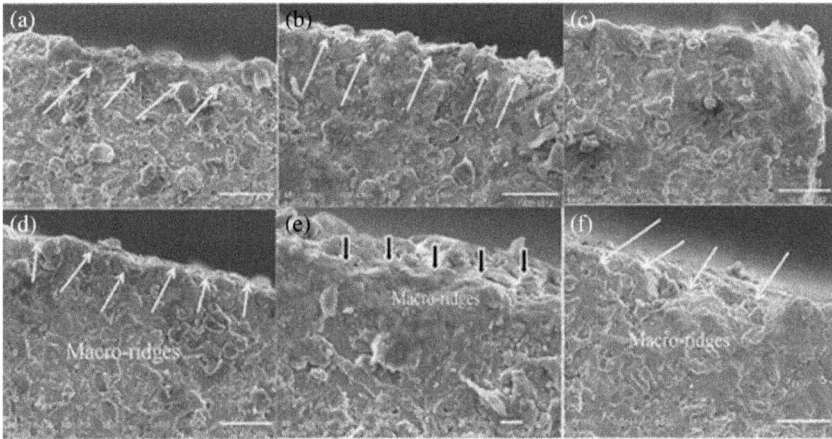

Figure 13.3 Minor hillocks, valleys, and macroridges at PC =160A, POFT = 26 µs, and POT = 0.9 µs.

at intermediate PC (160 A) and POT (0.9 µs). As seen in Figure 13.3(d)–(f), WEDMed may also include macroridges made by melted material. Low PC (120 A) and POT (0.7 µs) caused fewer craters and no microcracks, as observed in Figure 13.4(a)–(d). The work surface imposes less intense discharge due to low PC and POT [25]. See Figures 13.3 and 13.4.

13.3.2 Effect of WEDMed variables on heat-affected zone

Two layers, the heat-affected zone (HAZ) and the recast layer (RL), were deposited to the machined surface. WEDMed causes the material to change phase because of the extreme temperature variations in the work material. Rapid heating and quenching in the spark gap of the WEDM process generated the heat-affected zone [26]. As observed in Figure 13.5, the work material and wire electrode elements melted and resolidified at the heat-affected zone. Figure 13.5(b) and (d) shows the residuals of a spherical. As demonstrated in Figure 13.5(a), high PC (200 A) leads to the formation of cracks. The formation of a crack in the HAZ surface was produced when stress surpassed the material's ultimate tensile strength [27–29]. Due to the limited thermal conductivity of pure titanium, a feeble annealed layer can be seen in Figure 13.5(d). It might be due to the slower heat dissipation and quick dielectric cooling. It's remarkable to note that several spherical deposits were also observed in Figure 13.5(d)–(f). An O_2-rich layer was observed adjacent to the spherical deposits. The bull's-eye mark was used to describe these deposit characteristics.

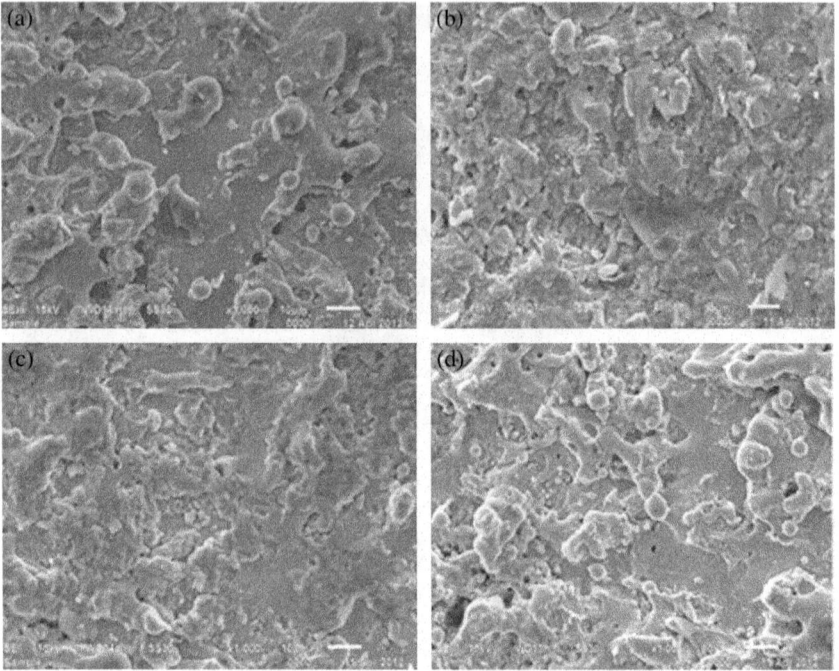

Figure 13.4 Micrographs with few numbers of craters and no cracks observed at lower PC =120 A, POT= 0.7 μs, and POFT = 38 μs and SR = 2.15 μm, 2.23 μm, 2.48 μm, 2.28 μm, 2.42 μm, and 2.35 μm.

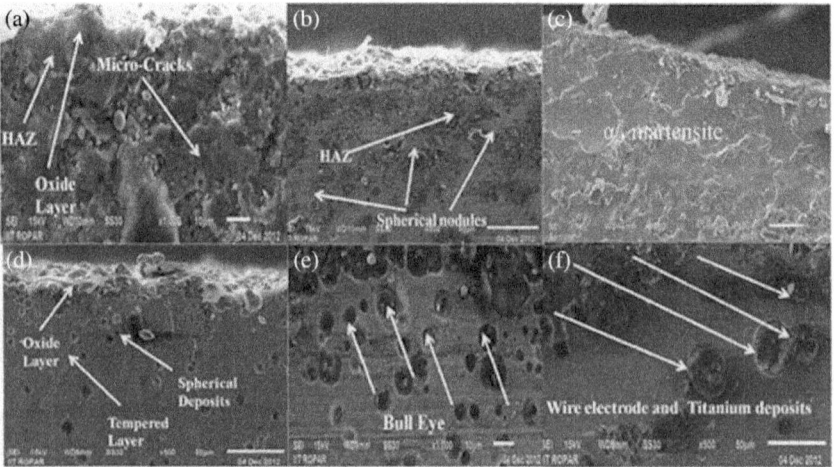

Figure 13.5 SEM micrographs of heat-affected zone (HAZ) at higher POT = 0.9 μs, POFT = 17 μs, and PC=200 A.

13.3.4 Effect of WEDMed variables on a recast layer (RL)

The recast layer was formed once molten material was swept across the cross section of the work sample and is not removed by dielectric fluid [30–32]. It is tremendously problematic to remove, and its appearance was observed at various magnification levels. The breadth of the recast layer of the WEDMed surface was increased due to an increase in PC, POT, and a corresponding drop in POFT, as shown in Figure 13.6(a)–(h). Its topography appears like a wave-like pattern and is nonuniform in appearance. Some microcracks were also observed trenchantly into the recast layer, which cleaves into the adjoining matrix phase, as illustrated in Figure 13.6(d). Dendrites and sublayers with separate orientations are labeled A, B, C, and D in Figure 13.6(f). The variation of average thickness was observed between 6 and 58 μm.

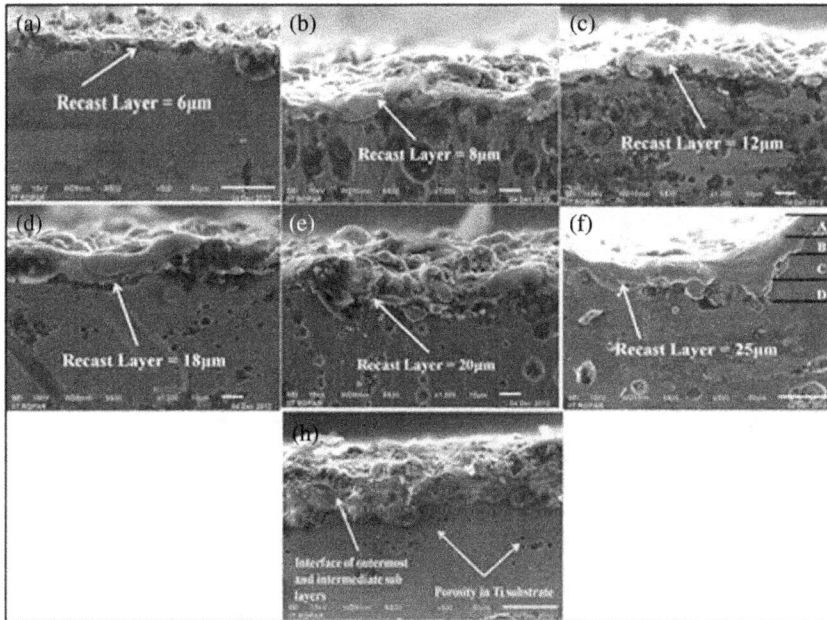

Figure 13.6 Recast layer at different parametric conditions: (a) POT = 0.7 μs, POFT = 38 μs, and PC = 120 A; (b) POT = 0.9 μs, POFT = 26 μs, and PC= 160 A; (c) POT = 0.9 μs, POFT = 17 μs, and PC = 160A; (d) POT = 1.1 μs, POFT = 26 μs, and PC = 160 A; (e) POT = 1.1 μs, POFT = 17 μs, and PC = 200 A; (f) POT = 1.1 μs, POFT = 17 μs, and PC = 200 A (Marker A: outmost sublayer, B: intermediate sublayer, C: innermost recast zone, D: matrix zone; (g) recast layer with dendrite structure; (h) porosity in Ti substrate.

13.3.5 Effect of WEDMed variables on the microcrack formation

Microcrack development was influenced not only by machining factors but also by material properties [33]. Machined samples with subsurface cracks revealed that the material was amorphous. Microcrack growth is often followed by fast quenching and heating by the dielectric. Cracks at different settings of PC and POT were observed on the subsurface, as in Figure 13.7(a)–(f). Due to the free movement of debris, there was a higher density of microcracks [34]. The microcracks demonstrated that, as the level of POT and PC increased, the severity of surface cracks started to rise here too. To determine the severity of cracking, a "surface crack size density" is calculated by multiplying the total length of cracks (m) by the unit area (m²). This observation is supported by the fact that PC and POT increase discharge energy and impulsive force, resulting in larger cracks and deeper craters.

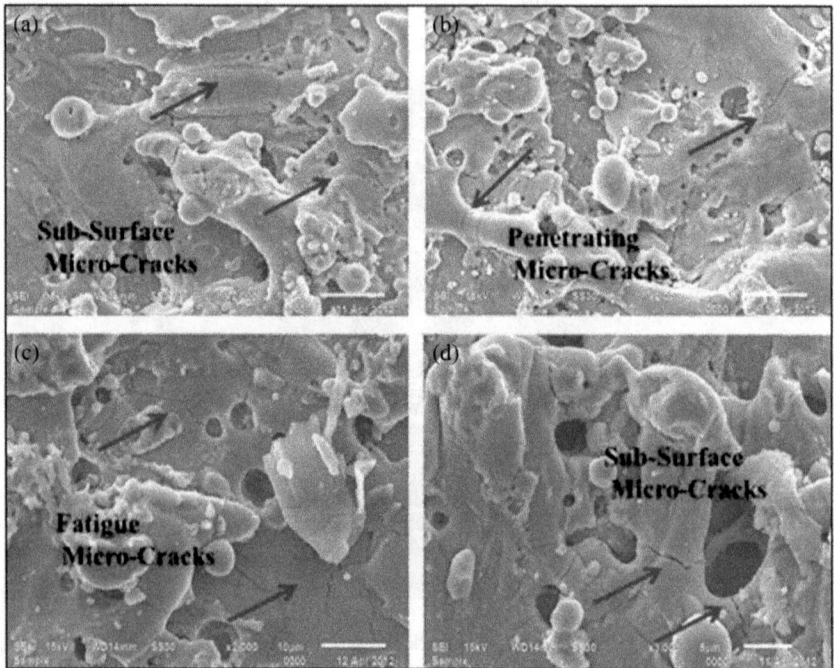

Figure 13.7 Micrograph of subsurface microcracks and fatigue cracks at Run nos. 20, 21, 31.

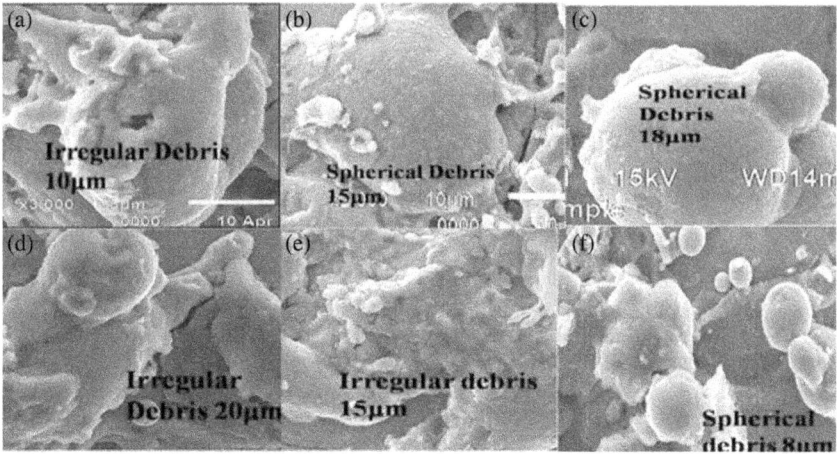

Figure 13.8 Debris collected during WEDM of pure titanium using deionized water: (a) MRM: spalling, (b) MRM: melting, (c) MRM: melting, (d) MRM: spalling, (e) MRM: spalling, (f) MRM: melting.

13.3.6 Effect of WEDMed variables on spalling

The material removal mechanism (MRM) in the WEDM process spalls, causing a small amount of material to be extracted from the base material [35]. The most common cause of spalling is the formation of bigger microcracks during WEDM. A debris study, as shown in Figure 13.8, reveals the spalling action Figure 13.8(a)–(f). These particles are uneven in shape, with rounded, spherical, and sharp edges, indicating that spalling is used to remove material.

13.3.7 Effect of WEDMed variables on analysis of wire electrode surface topography

In the present research, two forms of wire rupture were observed. These could occur as a result of the high PC and spark rate. The micrographs of the worn-out wire electrode are shown in Figure 13.9(a)–(h). On the deteriorated wire's surface, craters and residual debris were visible. EDX was used to examine the residuals of Cu, C, O, and Ti elements in the brass wire. Arcing was caused by debris getting stuck in the gap and not being flushed out effectively by the dielectric [36–38].

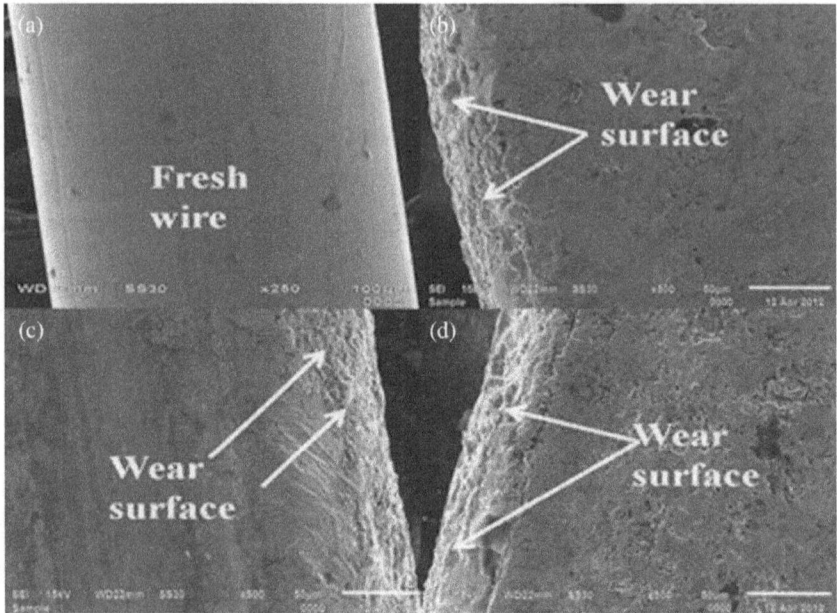

Figure 13.9 Micrographs (250× and 500×) of the wear wire electrode surface for different parametric conditions: (a) POT = 0.7 µs, PC = 120 A, (b) POT = 0.9 µs, PC = 160A, (c) POT = 0.9 µs, PC = 160 A, (d) POT= 1.1 µs, PC= 160 A.

13.3.8 Analysis of energy dispersive X-ray analysis (EDX) and X-ray diffraction analysis (XRD)

The composition of migratory elements on the workpiece and electrode surfaces was investigated using energy dispersive X-ray (EDX) [39–40]. The greater peak spectrum suggests that the work sample contains more focused elements. The patterns of EDX of subsurfaces with a 3 KV accelerating voltage are shown in Figure 13.10(a)–(d). Significant amounts of elements migrated to the surface of the work samples, according to the observations. Using X'Pert High Score plus 2.0, researchers were able to recognize the key phases by observing the XRD patterns. The phases were determined utilizing the PDF-2 database. On the machined subsurface were compounds such as titanium dioxide (rutile), ilmenite, copper titanium dioxide, zinc titanium carbide, titanium zinc carbide, titanium carbide, titanium carbide, as shown in Figure 13.11(a)–(f). Peaks on the 2Θ scale with values of 35.45, 45.67, 65.70, 85.07, and 105.69 (Cu K) correspondingly were used to identify the phases.

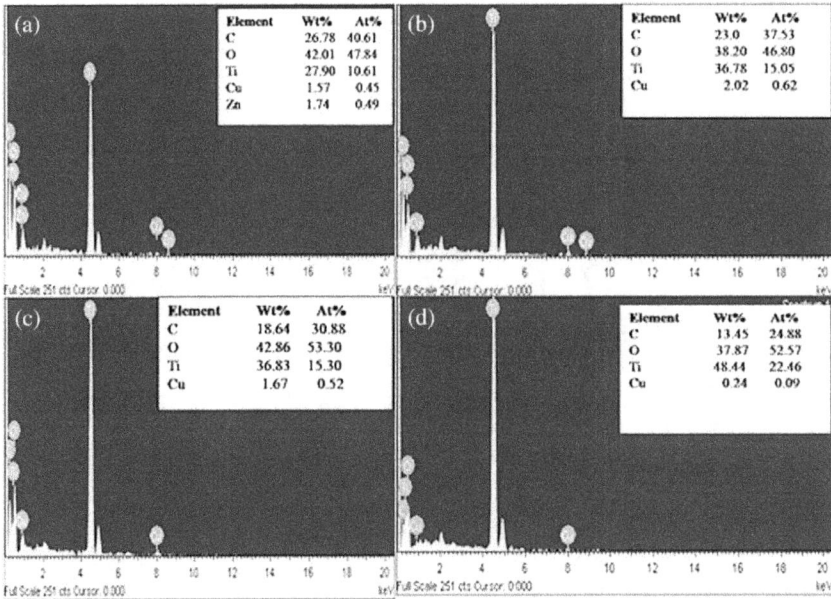

Element	Wt%	At%
C	26.78	40.61
O	42.01	47.84
Ti	27.90	10.61
Cu	1.57	0.45
Zn	1.74	0.49

Element	Wt%	At%
C	23.0	37.53
O	38.20	46.80
Ti	36.78	15.05
Cu	2.02	0.62

Element	Wt%	At%
C	18.64	30.88
O	42.86	53.30
Ti	36.83	15.30
Cu	1.67	0.52

Element	Wt%	At%
C	13.45	24.88
O	37.87	52.57
Ti	48.44	22.46
Cu	0.24	0.09

Figure 13.10 EDX analysis of migrated elements at different experiment nos. vs. atomic wt.%.

13.4 CONCLUSIONS

In conclusion, for the WEDMed process, an RSM-based BBD approach was used to simultaneously distinguish surface and subsurface abnormalities such as crack density and the RLT of titanium grade-2 by process parameters, i.e. POT, POFT, PC, and SGV. The following are the key findings:

1. The surface morphology of machined samples has deteriorated with deep and wide overlapping craters, pockmarks, globules of debris, and microcracks with high POT and PC.
2. Thermal and surface tension processes adjacent to the heat-affected zone produced the spherical nodules to grow. With higher PC, POT, and low POFT, the recast layer was observed. The recast layer was formed by the rapid quenching and heating of the dielectric, and microcracks were also observed in the recast layer due to high PC.
3. Rounded, spherical, and sharp edged debris formed due to spalling. The subsurface became very hard and brittle due to carbon content increases, reinforcing the surface cracking. The WEDMed surface was observed with migrated elements that are shown on the EDX peaks pattern with atomic and weight percentages.

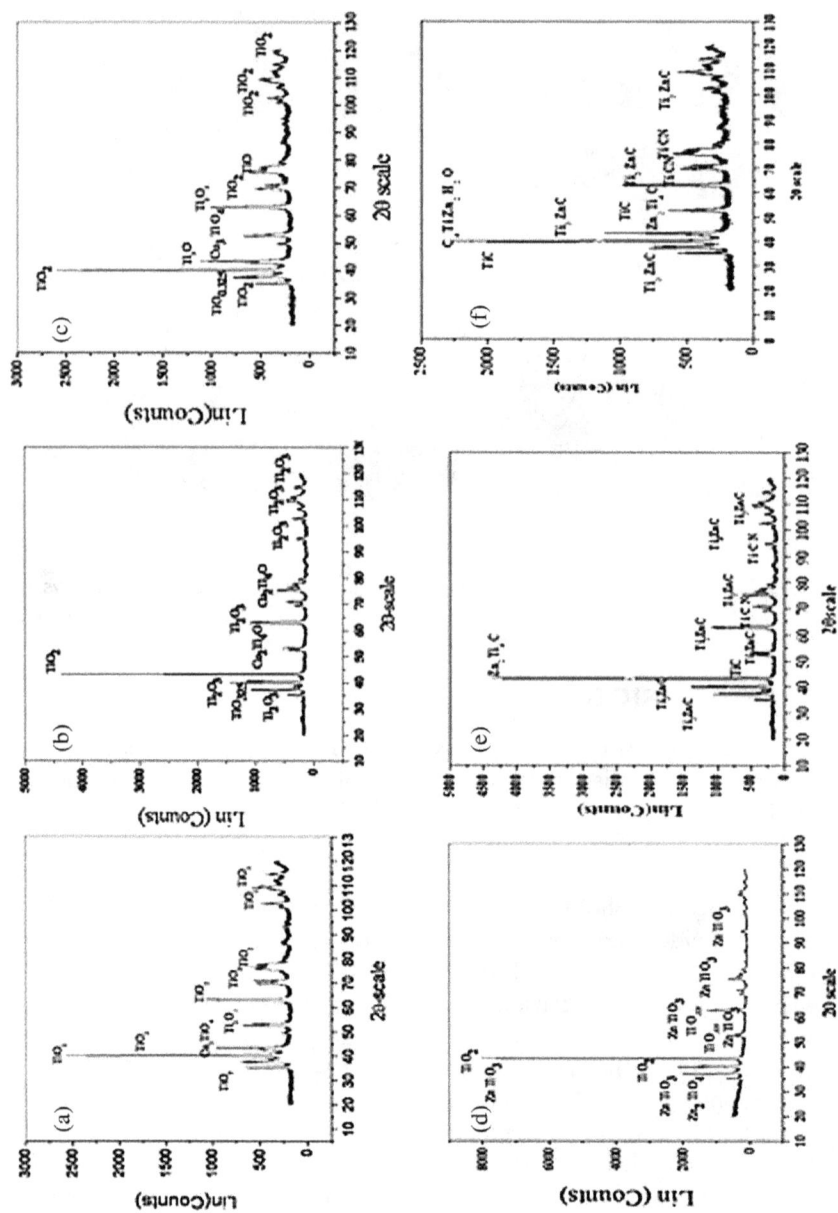

Figure 13.11 XRD phase pattern analysis of the machined surface.

4. Through XRD analysis, the chemical compounds and phases of titanium dioxide, ilmenite, copper titanium tetradioxide, titanium, carbide (TiC) have been observed.
5. The crack density and RLT rise as the POT increases (137 and 43%, respectively). Higher POT on machined CP-Ti increases crack density while also deteriorating the surface texture with debris lumps. Furthermore, the crack density and RLT diminish as the POFT is extended from 17 to 38 µs (37%, 8%).

13.5 FUTURE SCOPE

The present work is an attempt toward the investigation of subsurface characteristics of titanium grade-2 using WEDMed. The research work could be further extended over a variety of some more advanced materials such as composites, shape memory alloys, and alloy steels.

REFERENCES

1. Elias, C., Lima, J., Valiev, R., Meyers, M. (2008) Biomedical applications of titanium and its alloys, The Journal of The Minerals, Metals & Materials Society, 60(3):46–49.
2. Long, M., Rack, H. (1998) Titanium alloys in total joint replacement-a materials science perspective, Biomaterials, 19(18):1621–1639.
3. Budinski, K.G. (1991) Tribological properties of titanium alloys, Wear, 151(2):203–217.
4. Revankar, G.D., Shetty, R., Rao, S.S., Gaitonde, V.N. (2017) Wear resistance enhancement of titanium alloy (Ti-6Al-4V) by ball burnishing process, Journal of Material Research Technology, 6(1):13–32.
5. Isil, Yemisci et al. (2019) Experimentation and analysis of powder injection molded Ti10Nb10Zr alloy: a promising candidate for electrochemical and biomedical application, Journal of Material Research Technology, 8(6):5233–5245.
6. Prakash, C., Kansal, H.K., Pabla, B., Puri, S., Aggarwal, A. (2016) Electric discharge machining-a potential choice for surface modification of metallic implants for orthopedic applications: a review, Proceedings of the Institution of Mechanical Engineers, Part B: Journal of Engineering Manufacture, 230(2):331–353.
7. Calin, M., Gebert, A., Ghinea, A.C., Gostin, P.F., Abdi, S., Mickel, C., Eckert J (2013) Designing biocompatible Ti-based metallic glasses for implant applications, Materials Science and Engineering C, 33(2):875–883.
8. Kuriakose, S., Shunmugam, M.S. (2005) Multiobjective optimization of wire electro- discharge machining process by non-dominated sorting genetic algorithm, Journal of Material Processing Technology, 170 (1-2):133–141.
9. Sarkar, S., Mitra, S., Bhattacharyya, B. (2005) Parametric analysis and optimization of wire electrical discharge machining of γ-titanium aluminide alloy, Journal of Material Processing Technology, 159(3):286–294.

10. Sarkar, S, Mitra, S., Bhattachyraya, B. (2006) Parametric optimization of wire electric discharge machining of γ titanium alumnide alloy through an artificial neural network model, International Journal of Advance Manufacturing Technology, 27:501–508.

11. Sarkar, S., Sekh, M., Mitra, S., Bhattacharyya, B. (2008) Modeling and optimization of wire electrical discharge machining of γ-TiAl in trim cutting operation, Journal of Material Processing Technology, 205:376–387.

12. Hseigh, S.F., Chen, S.L., Lin, M.H. (2009) The machining characteristics and shape recovery ability of Ti Ni X(X = Zr, Cr) ternary shape memory alloys using WEDM, International Journal Machine Tool and Manufacture, 49:509–514.

13. Hasçalık, A., Çaydaş, U. (2007) Electrical discharge machining of titanium alloy (Ti-6Al-4V), Applied Surface Science, 253(22):9007–9016.

14. Huang, C.A., Hsu, F.Y., Yao, S.J. (2004) Microstructure analysis of the martenstic stainless steel surface fine-cut by the wire electrode discharge machining (WEDM), Materials Science and Engineering A, 371:119–126.

15. Kung, Yuan, Kuang, Chiang, Ta-Ko (2008) Modeling and analysis of machinability evaluation in the wire electric discharge machining (WEDM) process of Aluminum oxide-based ceramics, Material and Manufacture Processes, 23:241–250.

16. Liao, Y.S., Yu, Y.P. (2003) Study of specific discharge energy in WEDM and its application, International Journal Machine Tool and Manufacture, 44(12-13):1373–1380.

17. Porous, D., Zaboruski, S. (2009) Semi empirical model of efficiency of wire electric discharge machining of hard to machine materials, Journal of Material Processing Technology, 209(3): 1247–1253.

18. Hseigh, S., F, Chen, S.L., Lin, M.H. (2009) The machining characteristics and shape recovery ability of Ti Ni X(X= Zr, Cr) ternary shape memory alloys using WEDM, International Journal Machine Tool and Manufacture, 49:509–514.

19. Garg, M.P., Jain, A., Bhushan, G. (2013) Modeling and multi-objective optimization of process parameters of wire electrical discharge machining using non-dominated sorting genetic algorithm-II, Proceedings of the Institution of Mechanical Engineers, Part B: Journal of Engineering Manufacture, 226:1986–2001.

20. Aspinwall, Soo, D.K, Berrisford, S.L., Walder, G. (2008) Work piece surface roughness and integrity after WEDM of Ti-6Al-4V and Inconel-718 using minimum damage generator technology, CIRP- Annals of Manufacturing Technology, 57:187–190.

21. Kumar, A., Kumar, V., Kumar, J. (2013a) Microstructure analysis and material transformation of pure titanium and tool wear surface after wire electric discharge machining, Machining Science and Technology, 18(1):47–77.

22. Kumar, A, Kumar, V, Kumar, J. (2012) An investigation into machining characteristics of commercially pure Titanium (Grade-2) using CNC WEDM, Applied Mechanics and Materials, 159:56–68.

23. Ramakrishnan, R., Karunamoorthy, L. (2006) Multiresponse optimization of wire EDM operations using robust design of experiments, International Journal of Advance Manufacturing Technology, 29:105–112.
24. Govindan, P., Joshi, S.S.(2012) Analysis of micro-cracks on machined surfaces in dry electrical discharge machining, Journal of Manufacturing Processes, 14:277–288.
25. Ekmekci, B. (2009) White layer composition, heat treatment, and crack formation in electric discharge machining process, Metallurgical and Materials Transactions B, 40(1):70–81.
26. Puri, A.B., Bhattacharya, B. (2005) Modeling and analysis of white layer depth in wire cut EDM process through response surface methodology, International Journal of Advance Manufacturing Technology, 25:301–307.
27. Kumar, A., Kumar, V., Kumar, J. (2013b) Multi-response optimization of process parameters based on response surface methodology for pure titanium using WEDM process. International Journal of Advance Manufacturing Technology, 68(9-11):2645–2668.
28. Prakash, C., Kansal, H.K., Pabla, B., Puri, S. and Aggarwal, A. (2016) Electric discharge machining-a potential choice for surface modification of metallic implants for orthopedic applications: a review, Proceedings of the Institution of Mechanical Engineers, Part B: Journal of Engineering Manufacture, 230(2):331–353.
29. Zhang, Yanzhen, Liu, Yonghong, Ji, Renjie, Cai, Baoping (2011) Study of the recast layer of a surface machined by sinking electrical discharge machining using water-in-oil emulsion as dielectric. Applied Surface Science, 257:5989–5997.
30. Kuruvila, N., H.V. Ravindra (2011) Parametric influence and optimization of wire EDM of Hot die steel, Machining Science and Technology, 15:47–75.
31. Shahali, Hesam, Yazdi, Soleymani, Reza, M., Mohammadi, Aminollah, Limanian, Ehsan (2012) Optimization of surface roughness and thickness of white layer in wire electrical discharge machining of DIN 1.4542 stainless steel using micro-genetic algorithm and signal to noise ration techniques, Proceedings of the Institution of Mechanical Engineers, Part B: Journal of Engineering Manufacture, 5:803–812.
32. Huang, C.A., Tu, G.C., Yao, H.T., Kuo, H.H. (2004) Characteristics of the rough cut-surface of quenched and tempered maternsitic stainless steel using wire electrical discharge machining, Metallurgical and Materials Transactions A, 35A:1351–1357.
33. Govindan, P., Joshi, Suhas S. (2012) Analysis of micro-cracks on machined surfaces in dry electrical discharge machining, Journal of Manufacturing Processes, 14:277–288.
34. Tai, Tzu-Yao, Lu, S.J., Chen, Y.H. (2011) Surface crack susceptibility of electro discharge-machined steel surfaces. International Journal of Advance Manufacturing Technology, 57:983–989.
35. Tiley, J.T., Searles (2004) Quantification of microstructure features in Alpha/Beta Titanium alloys, Materials Science and Engineering A, 372:191–198.

36. Puri, A.B., Bhattacharyya, B. (2003) An analysis and optimization of the geometrical inaccuracy due to wire lag phenomenon in WEDM, International Journal Machine Tool and Manufacture, 43:151–159.
37. Tosun, Nihat, Cogun, Can. (2003) An investigation on wire wear in WEDM, Journal of Material Processing Technology, 2003; 134:273–278.
38. Tosun, N., Cogun, C., Inan, A. (2003) The effect of cutting parameters on workpiece surface roughness in Wire EDM, Machining Science and Technology, (2):209–219.
39. Hascalik, Ahmet, Caydas, Ulas. (2004) Experimental study of wire electrical discharge machining of AISID5 tool steel, Journal of Material Processing Technology, 148:362–367.
40. Boyer, R.G., Welsch, E.W., Collings (1994) Materials Property Handbook: Titanium Alloys. Materials Park, OH: ASM International.

Chapter 14

Additive manufacturing process based **EOQ** model under the effect of pandemic **COVID-19** on non-instantaneous deteriorating items with price dependent demand

Sanjay Sharma, Anand Tyagi, Sachin Kumar, and Priyanka Kaushik

14.1 INTRODUCTION

COVID-19 is still having a major impact on almost all types of businesses, though it is most noticeable in small cities in low-income countries such as India, Pakistan, Afghanistan, Bangladesh, Nepal, and Bhutan. One more important thing is that this situation of COVID-19 teaches us a lesson that we must always be ready for ups and downs in business and that we must plan something good effectively to face such a challenge as a pandemic. Products like vegetables, milk products, etc., whose shelf life is low and whose spoilage rate is high, are generally collected from small farmers in different villages, so the effect of COVID-19 has initially affected these products because of the prolonged lockdown. Even the process of collecting these products from different local areas has become tougher. So there is a gap between the collection and distribution of the items, and in this gap period, the product starts to deteriorate.

In this study, we discuss and develop a model for low-life-cycle products such as vegetables, fish, and meat. As various research has already been done to maximize total profit, we have developed this model under the conditions

DOI: 10.1201/9781003360001-14

of COVID-19. Various parameters are discussed in the study, and the effect of COVID-19 on these parameters is also examined.

Goyal et al. [1] presented their inventory model under the assumption that the order placing policy needs to be based on stock level and price under partially backlogged conditions. To reduce the amount of deterioration, Singh et al. [2] developed an inventory model on economic order quantity (EOQ) for deteriorating items making use of preservation technology.

Tiwari et al. [3] developed a model by considering the items whose life rate is high with selling-price-dependent demand under a fuzzy environment and where green technology is used to reduce carbon emissions, affecting the total profit. Some of the work was also developed by Manna et al. [4], Mukhopadhyay and Goswami [5], Pasandideh et al. [6], Nobil et al. [7], Cardenas Barron et al. [8], and many others to maximize the total profit. Deterioration is the process of reducing quality from higher to lower levels. It suggests change, obsolescence, spoilage, loss of utility, or loss of minor estimation of merchandise, resulting in the loss of its value. The deterioration rate is not almost insignificant for items like toys, glassware, hardware, and steel, but it is much more effective for items like vegetables, natural products, drugs, unpredictable fruits, blood, and innovative items

Deteriorated items have different physical highlights in which a large portion of products undergo deterioration over time and have different aspects as they do. This phenomenon is especially valid for items like natural products such as vegetables. Some specific items continuously deteriorate in the mortification process, whereas consumption rates are very large for other products like gasoline, alcohol, and turpentine. Hence deterioration assumes a vital role in the inventory model that should not be ignored. Also, in real life, deterioration is a common phenomenon. In fact, research on deteriorating product has been going on continuously since 1963. Share and Schrader [9] developed a model with a constant deterioration rate, which was validated with no shortages. A rapid change in deterioration like the exponential function was introduced by Liao [10] in 2008. Tripathy and Pandey [11] considered an inventory model with deteriorating items under a trade credit policy.

By assuming the deterioration as constant, Agarwal and Jaggi [12] developed a model. In the last two decades, many papers have been published in which demand is assumed to be price dependent. Chen and Chen [13] proposed an inventory model with finitely deteriorating items over a finite horizon. Chang et al. [14] developed a model for deteriorating items with a partial backlog for the retailer's optimal pricing and lot-sizing policies. Dye [15] considered the logic over an infinite time horizon with allowable back-ordering. Similar papers were developed by Panda et al. [16], Tsao and Sheen [17], Lin et al. [18], Dye and Ouyang [19], and Wang and Lin [20].

For the items, an inventory model, under joint optimal pricing and inventory control for deteriorating products under inflation and customer returns, was proposed by Ghoreishi et al. [21], where a joint pricing and replenishment policy with credibility constraint for non-instantaneous deteriorating items with imprecise deterioration items was developed by Soni and Patel [22], Srivastava and Gupta [23], Wang and Huang [24], Bai et al. [25], Bhunia et al. [26], Rabbani et al. [27], Indrajit Singha et al. [28, 29]. Recently, one work was developed by Rahamann and Duary [30]. They developed inventory control models for -instantaneous deteriorating items.

Under the environment of the COVID-19 pandemic, Kumar et al. [31] developed an inventory model by considering the learning effect on decaying items under preservation technology during the pandemic. Indrajit Singha et al. [32] presented an EOQ model of selling-price-dependent demand for non-instantaneous deteriorating items.

By considering the post effect of COVID-19, Mashud et al. [33] proposed a resilient hybrid payment supply chain inventory model where a strong study based on post-COVID-19 recovery was established. Yadav et al. [34] presented their model based on colony optimization and solved the effect of the COVID-19 pandemic on blood supply chain inventory management.

Various studies are being conducted in response to the COVID-19 pandemic to maximize and properly utilize the available inventory. In the present study, a model is developed for non-instantaneous deterioration products with a price and stock-dependent demand rate and shortages under partially backlogged conditions by making different combinations of the preceding system parameters. The study's main objective is to optimize by minimizing the total average cost per unit time.

14.2 ASSUMPTIONS AND NOTATIONS

The following assumptions are considered throughout the paper:

1. The market demand rate is dependent on the market price (p) of an item of the non-instantaneous deteriorating products, and it is of the form $D(s) = ap^{-b}$, a, $b > 0$.
2. As the items are non-instantaneously deteriorating in nature, the rate is taken to be constant throughout the cycle.
3. As the model is also based on deterioration, we assume that there is no deterioration between the time interval $[0, t_1]$ but it is also assumed that items start deteriorating after that period at a rate θ.
4. Replenishment of decaying or damaged items is not acceptable during the deterioration time.
5. The lead time is assumed to be zero.
6. Shortages are allowed and partially backlogged.

The notations used in the present model are as follows:

$I(t)$ on-hand inventory
D   demand size during the fixed cycle time T
a, b   demand coefficient
t_1 time during which there is no deterioration
h   holding cost (in ₹/unit)
θ   deterioration coefficient, $0 \leq \theta \leq 1$
t_2 The time at which the inventory level reaches to zero
ξ   lost sale cost (in ₹/unit)
R   ordering-cost (in ₹/order)
p   purchasing cost (in ₹/unit)
T   length of cycle
Q   ordering quantity
q_i on-hand inventory
q_s back-order inventory
r   shortage cost (in ₹/unit)
η   rate of backlogging
S   selling price
$I_1(t)$ inventory level, at any time t, during $[0, t_1]$
$I_2(t)$ inventory level, at any time t, during $[t_1, t_2]$
$I_3(t)$ inventory during the shortage period $[t_2, T]$
$T(v)$ total average inventory cost (in ₹)

14.3 MATHEMATICAL FORMULATION OF THE MODEL

In the model formulation in this section, we describe the development of an economic order quantity model of non-instantaneous deteriorating products that starts after some time interval. It is assumed that the experiment is started with q_i units of on-hand inventory and q_s units of back-order inventory, and it is assumed to be at the beginning of each cycle. Let T be the length of each cycle. Assume that, in the time interval $[0, t_1]$, there will be no deterioration of the product. The deterioration starts at $t = t_1$ and at $t = t_2$, there will no inventory. Hence for $t > t_2$, shortages appear, which happen up to the end of the cycle. To fulfill this shortage, partially backlogs are taken with a rate η. Let $I_1(t)$ and $I_2(t)$ be the inventory levels at any time t in the time interval $[0, t_1]$ and $[t_1, t_2]$, respectively. The behavior of the model is shown in Figure 14.1.

Thus we have the following differential equations for the proposed inventory system:

$$\frac{dI_1(t)}{dt} = -\left[ap^{-b}\right] \qquad 0 \leq t \leq t_1 \tag{14.1}$$

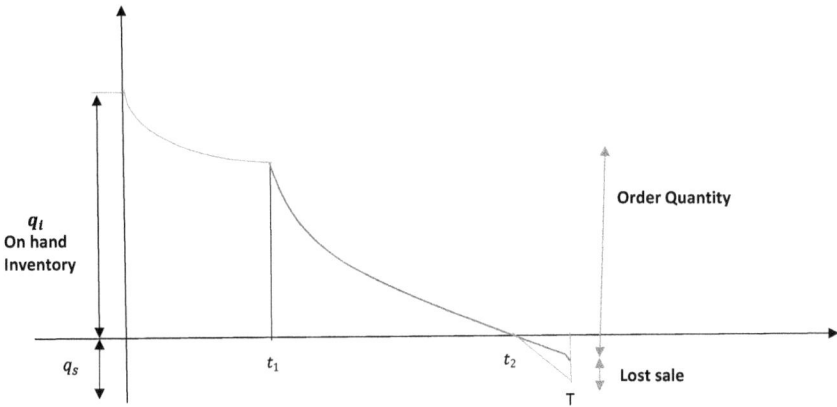

Figure 14.1 Graphical representation of inventory with respect to time.

$$\frac{dI_2(t)}{dt} + \theta I_2(t) = -[ap^{-b}] \quad t_1 \leq t \leq t_2 \tag{14.2}$$

$$\frac{dI_3(t)}{dt} = -[a] \quad t_2 \leq t \leq T \tag{14.3}$$

with boundary conditions:

$$I(t_2) = 0, I_3(T) = 0, I_1(t_1) = I_2(t_2) \tag{14.4}$$

Using these boundary conditions, the solutions of equations (14.1), (14.2), and (14.3) are

$$I_1(t) = ap^{-b}(t_1 - t) \tag{14.5}$$

$$I_2(t) = ap^{\;b}(t_2 - t) \tag{14.6}$$

$$I_3(t) = a(t_2 - t) \tag{14.7}$$

Using the initial condition $I_1(0) = q_i$, we have

$$q_i = ap^{-b}(t_1) \tag{14.8}$$

where the total average cost per unit time for the model during a complete cycle of time is given by

$$(v) = \frac{1}{T}\begin{bmatrix} \text{buying cost} + \text{inventory holding cost} + \\ \text{shortage cost} + \text{lost sale cost} + \text{ordering cost} \end{bmatrix} \quad (14.9)$$

Now the calculation of different costs associated with inventory is as follows:

1. Buying cost (BC): As total ordering quantity Q is the sum of on-hand inventory and back-order inventory, therefore $Q = q_i + q_s$.

   Therefore, the total buying cost can be formulated as

   $$\text{Buying cost (BC)} = Qp = [q_i + q_s]$$

   where $q_i = ap^{-b}(t_1)$

   and $q_s = \int_{t_2}^{T} [\, ap^{-b}\theta] \, dt = ap^{-b}[T - t_2]$

   $$\text{Buying cos t(BC)} = ap^{-b}[(t_1)a + p^{-b}[T - t_2] \quad (14.10)$$

2. Inventory holding cost (HC): Inventory holding cost is the cost to maintain inventory in the stock.

   $$\text{Holding cos t} = h\left[\int_{0}^{t_1} I_1(t)dt + \int_{t_1}^{t_2} I_2(t)dt\right] \quad (14.11)$$

3. Shortage cost (SC): This cost is associated with the cost during a stock-out situation; hence the shortage cost is given by

   $$\text{Shortage cos t(SC)} = \int_{t_2}^{T} [ap^{-b}\theta]dt = rap^{-b}[T - t_2] \quad (14.12)$$

4. Lost sale cost (LSC): This cost is associated with the inventory items we have lost during shortages.

   $$\text{Lost} - \text{sale cos t(LSC)}\int_{t_2}^{T} [[1 - \theta)\xi \, ap^{-b}\theta]dt = [(1 - \theta)\xi ap^{-b}\theta(T - t_2)] \quad (14.13)$$

5. Ordering cost (OC): OC = R   (14.14)

14.4 NUMERICAL EXAMPLES

Once we formulated some models, we took the following data to validate them: $a = 1500$, $b = 1.3$, $s = 5$, $t_1 = 3$, $\theta = 0.02$, $\xi = 2$, $h = 0.2$, $R = 190$, $\eta = 0.6$, $r = 7$, $p = 3$. The values of the different parameters considered here are realistic, though they are not taken from any case study. By using Mathematica 11.1 software, we get the unique $v = 9.1011$ and $Q = 514.091$ with optimum total average cost $(v) = ₹\,258.123$.

14.5 SENSITIVITY ANALYSIS

From the numerical example just mentioned, the following sensitivity analysis (Table 14.1 and Figures 14.2–14.10) was carried out to reveal the following managerial insights by changing the parameters one at a time, keeping others unchanged.

Table 14.1 Sensitivity analysis for different parameters

b	v	$K(v)$	Q	t_1	v	$K(v)$	Q
2.12	11.2011	372.206	688.135	4	10.8834	284.105	526.331
2.18	11.2011	342.107	640.235	5	11.1864	277.265	510.142
2.28	11.2011	338.116	624.235	6	11.2423	261.897	509.211
2.40	11.2011	325.206	611.235	7	11.5550	251.205	503.312
2.45	11.2011	312.206	605.425	8	12.2351	246.105	498.121
2.50	11.2011	278.206	590.235	9	13.1516	239.125	479.331
S	v	$K(v)$	Q	θ	v	$K(v)$	Q
4.5	11.2011	355.394	656.298	0.001	13.9267	260.926	564.6993
4.8	11.2011	342.267	612.298	0.012	13.8267	280.926	554.6993
5.1	11.2011	334.794	610.298	0.015	13.1267	290.926	544.6993
5.4	11.2011	324.394	496.298	0.017	12.4267	320.926	534.6993
5.7	11.2011	312.394	456.298	0.018	11.4267	340.926	524.6993
6.0	11.2011	236.314	416.298	0.019	10.4267	345.926	514.6993
ξ	v	$K(v)$	Q	H	v	$K(v)$	Q
4.5	10.8037	272.324	499.484	0.22	12.3535	262.479	565.344
5.0	11.1037	280.217	501.484	0.29	12.1535	280.214	545.344
5.5	11.6037	287.324	520.487	0.40	11.3535	289.415	525.344
6.0	11.7837	290.312	530.123	0.45	11.1535	309.476	504.344
7.0	11.8137	296.389	543.429	0.47	10.2535	316.210	490.344
7.5	11.9137	308.328	550.418	0.50	10.1535	328.476	410.290
R	v	$K(v)$	Q	P	v	$K(v)$	Q
150	11.2011	280.812	515.061	2.5	11.5336	265.464	528.773
170	11.2011	281.712	515.061	3.0	11.1234	290.464	510.712

(continued)

Table 14.1 Cont.

R	v	K(v)	Q	P	v	K(v)	Q
190	11.2011	282.812	515.061	3.5	10.9336	310.124	507.512
210	11.2011	286.652	515.06	4.0	10.7132	320.461	499.290
230	11.2011	290.812	515.061	4.5	10.2136	330.463	450.190
250	11.2011	295.812	515.061	5.0	10.1316	333.461	410.760

r	v	K(v)	Q	λ	v	K(v)	Q
6.5	10.3845	263.259	483.976	0.3	11.5051	287.205	492.311
7.0	10.9845	273.259	493.367	0.35	11.3412	290.205	497.311
7.5	11.1245	283.429	498.956	0.4	11.1234	295.205	503.327
8.0	11.6845	284.659	510.166	0.45	10.9051	321.207	512.319
8.5	11.8845	285.159	415.976	0.5	10.7256	325.267	520.318
9.0	11.9845	286.259	525.123	0.55	10.6051	330.123	525.317

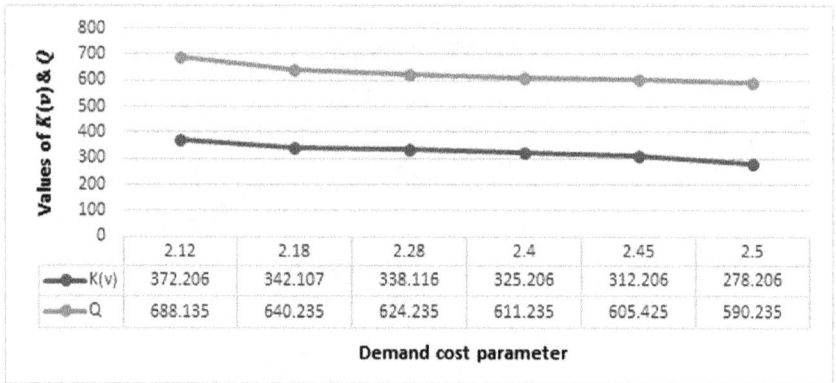

Demand cost parameter	2.12	2.18	2.28	2.4	2.45	2.5
K(v)	372.206	342.107	338.116	325.206	312.206	278.206
Q	688.135	640.235	624.235	611.235	605.425	590.235

Figure 14.2 Demand cost parameter (b) vs. K(v) and Q.

14.6 RESULT ANALYSIS

The following observations are made from Table 14.1 and from the corresponding figures (Figure 14.2–14.10):

1. It can be seen from Table 14.1 and its associated figures that, when demand parameter b increases, then a decrement is noted in the total average cost where ordering quantity Q is decreased under an increased value.
2. It is also discovered that, as the price of the item increases, both the ordering quantity Q and the total average cost K(v) decrease with the product constant. That is, the increase in price means an increase in the total profit.

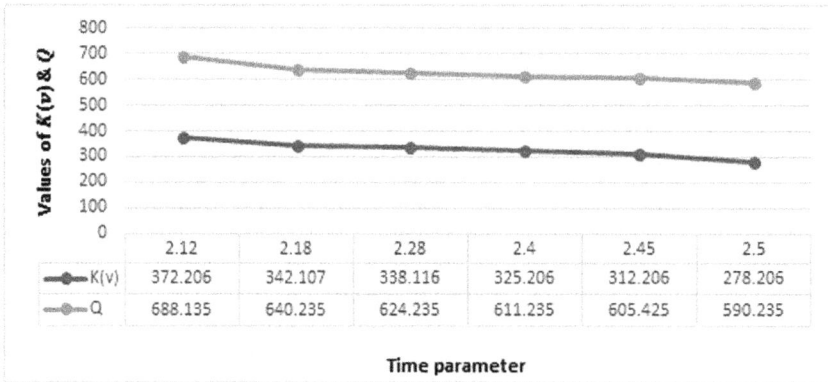

	2.12	2.18	2.28	2.4	2.45	2.5
K(v)	372.206	342.107	338.116	325.206	312.206	278.206
Q	688.135	640.235	624.235	611.235	605.425	590.235

Time parameter

Figure 14.3 Time parameter (b) vs. K(v) and Q.

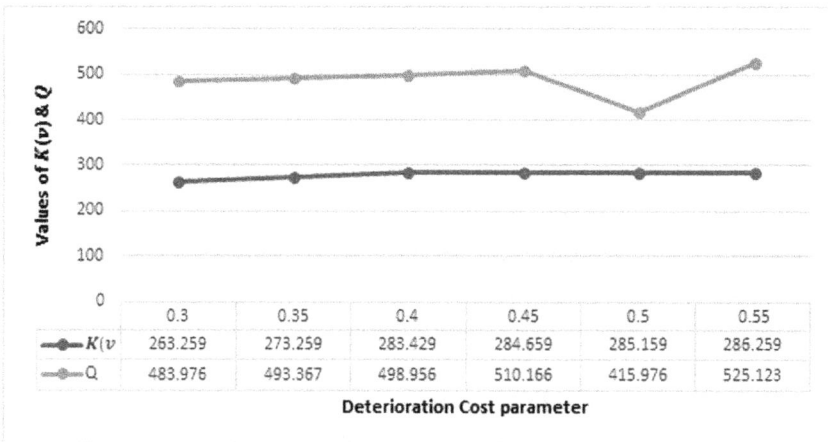

	0.3	0.35	0.4	0.45	0.5	0.55
K(v	263.259	273.259	283.429	284.659	285.159	286.259
Q	483.976	493.367	498.956	510.166	415.976	525.123

Deterioration Cost parameter

Figure 14.4 Deterioration coefficient (θ) vs. K(v) and Q.

3. When the deterioration parameter is increased, the total average cost K(v) of products such as vegetables, baked goods, fish, and meats increases. The time at which the inventory level becomes 0, the order quantity also decreases.
4. Table 14.1 also shows that, when the lost sale cost increases, the total average cost K(v) increases with an increase in order quantity, and ...
5. The increasing value holding cost h is accompanied by an increase in the total average cost K(v) per unit time of the products with a

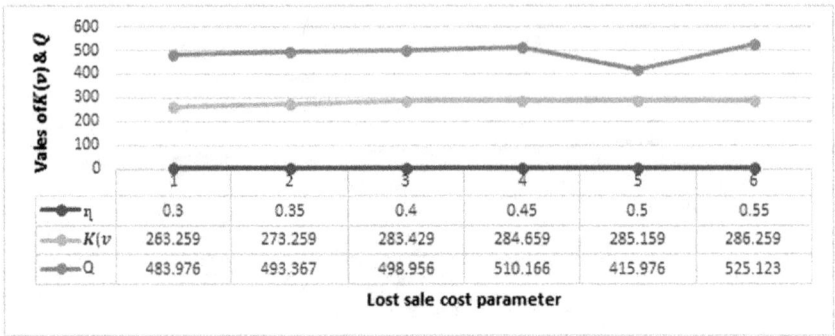

	1	2	3	4	5	6
η	0.3	0.35	0.4	0.45	0.5	0.55
K(v	263.259	273.259	283.429	284.659	285.159	286.259
Q	483.976	493.367	498.956	510.166	415.976	525.123

Lost sale cost parameter

Figure 14.5 Lost sale cost (ξ) vs. K(v) and Q.

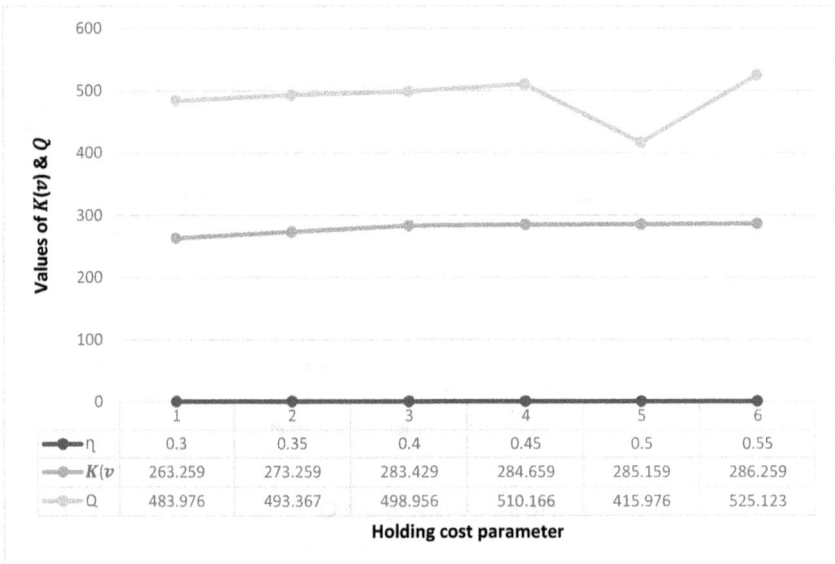

	1	2	3	4	5	6
η	0.3	0.35	0.4	0.45	0.5	0.55
K(v	263.259	273.259	283.429	284.659	285.159	286.259
Q	483.976	493.367	498.956	510.166	415.976	525.123

Holding cost parameter

Figure 14.6 Holding cost (h) vs. K(v) and Q.

decrease in ordering quantity Q and in the time at which the inventory level becomes available.
6. The total average cost K(v) was found to increase as the order quantity Q was reduced.
7. Under increased purchasing cost p values, the total average cost K(v) was found to be increased with a decrease in the ordering quantity Q and the time when there was no item in stock.

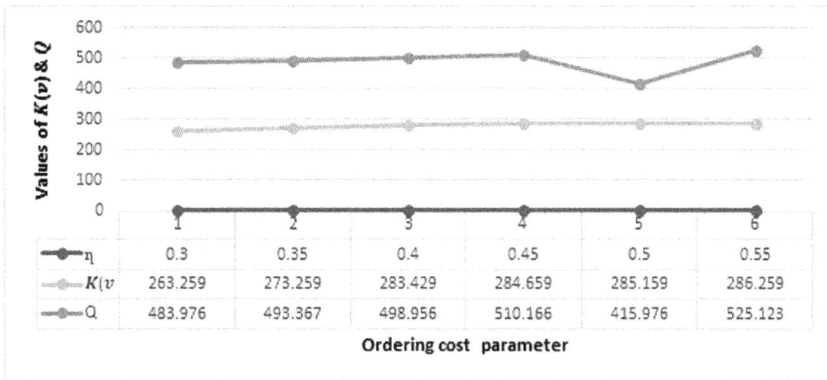

	1	2	3	4	5	6
η	0.3	0.35	0.4	0.45	0.5	0.55
K(v	263.259	273.259	283.429	284.659	285.159	286.259
Q	483.976	493.367	498.956	510.166	415.976	525.123

Ordering cost parameter

Figure 14.7 Ordering cost (R) vs. K(v) and Q.

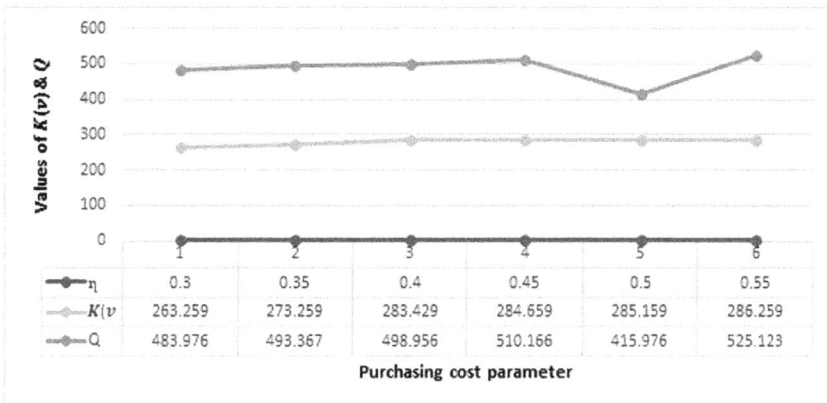

	1	2	3	4	5	6
η	0.3	0.35	0.4	0.45	0.5	0.55
K(v	263.259	273.259	283.429	284.659	285.159	286.259
Q	483.976	493.367	498.956	510.166	415.976	525.123

Purchasing cost parameter

Figure 14.8 Purchasing cost (p) vs. K(v) and Q.

8. When the shortage cost r increases, it is concluded that K(v) increases with the increase of the parameters Q and the time when there is no item in the stock.
9. It is positively seen that, if we make some increment in the backlogging parameter, the ordering quantity Q increases, and K(v) is found to decrease with a slight decrease in the value of the parameter.

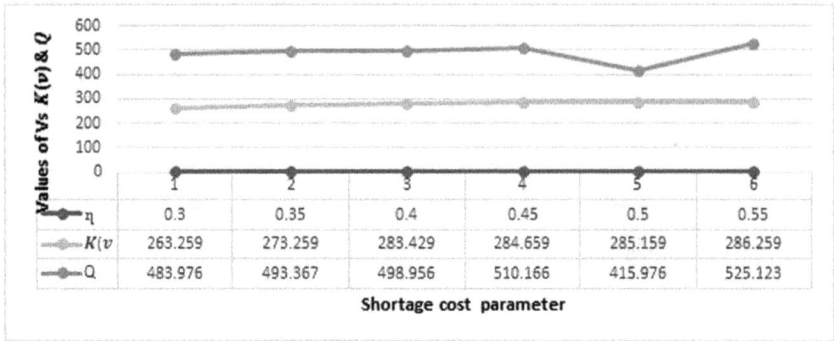

	1	2	3	4	5	6
η	0.3	0.35	0.4	0.45	0.5	0.55
K(v	263.259	273.259	283.429	284.659	285.159	286.259
Q	483.976	493.367	498.956	510.166	415.976	525.123

Shortage cost parameter

Figure 14.9 Shortage cost (r) vs. K(v) and Q.

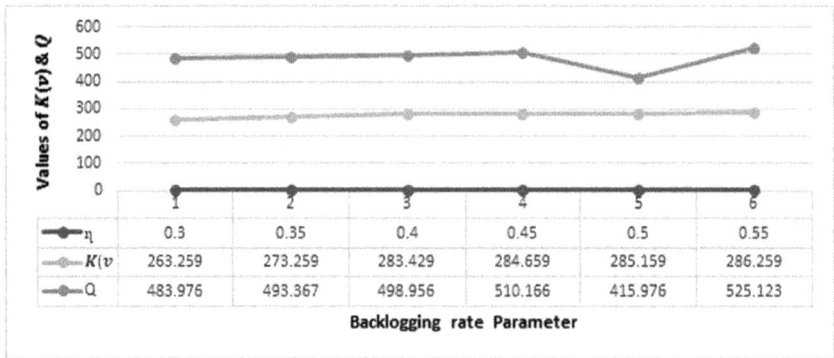

	1	2	3	4	5	6
η	0.3	0.35	0.4	0.45	0.5	0.55
K(v	263.259	273.259	283.429	284.659	285.159	286.259
Q	483.976	493.367	498.956	510.166	415.976	525.123

Backlogging rate Parameter

Figure 14.10 Backlogging rate (η) vs. K(v) and Q.

14.7 CONCLUSIONS

The current model is based on the additive manufacturing process in which the real-time situation of COVID-19 is considered and ways of facing COVID-19 are developed. Most countries in the world are affected by the COVID-19 pandemic and are facing the same problem of not satisfying their customers' demands because of factors that are directly responsible for proper inventory management. In particular, medium- and small-scale entrepreneurs are most affected by the pandemic. As this situation may be going on for a while, we must put all our efforts into making some decisions to make better arrangements for the proper distribution of goods and for meeting demands as soon as possible.

In the same way that the current study presents an inventory control model that is formulated and developed specifically for the current COVID-19 situation of the COVID-19 pandemic and focuses on items such as vegetables, bakery, fish, and meats that do not deteriorate instantly, we consider non-instantaneous decaying products. In the present model, a procedure for obtaining the solution is provided to get the optimum total average cost, and practical application examples support the application of the proposed model to support decision making.

It is observed that it is effective to take the demand as price and stock dependent. Also, it is observed in the study that it is more realistic to assume that we must monitor the price of the item during a stock-out situation carefully and that better planning during the shortage period can make a big difference in the profit.

Research along this line can be used to study the impact of time-varying deterioration on optimal policy.

14.8 STUDY'S FUTURE SCOPE

The present study may be further expanded by making some more changes in the main parameters of the study. For example, this model can also be developed assuming variable deterioration, trade credits, and some price discount for the on-hand inventory may also be assumed.

REFERENCES

[1] Goyal, A. K., Chauhan, A., & Singh, S. R. (2015). An EOQ inventory model with stock and selling price dependent demand rate, partial backlogging, and variable ordering cost. *International Journal of Agricultural and Statistical Sciences, 11*(2), 441–447.

[2] Singh, S., Khurana, D., & Tayal, S. (2016). An economic order quantity model for deteriorating products having stock dependent demand with trade credit period and preservation technology. *Uncertain Supply Chain Management, 4*(1), 29–42.

[3] Tiwari, S., Ahmed, W., & Sarkar, B. (2019). Sustainable ordering policies for non-instantaneous deteriorating items under carbon emission and multi-trade-credit-policies. *Journal of Cleaner Production, 240*, 118183.

[4] Manna, A. K., Dey, J. K., & Mondal, S. K. (2017). Imperfect production inventory model with production rate dependent defective rate and advertisement dependent demand. *Computers & Industrial Engineering, 104*, 9–22.

[5] Mukhopadhyay, A., & Goswami, A. (2017). An inventory model with shortages for imperfect items using substitution of two products. *International Journal of Operational Research, 30*(2), 193–219.

[6] Pasandideh, S. H. R., Niaki, S. T. A., Nobil, A. H., & Cárdenas-Barrón, L. E. (2015). A multiproduct single machine economic production quantity model for an imperfect production system under warehouse construction cost. *International Journal of Production Economics*, *169*, 203–214.

[7] Nobil, A. H., Sedigh, A. H. A., & Cárdenas-Barrón, L. E. (2016). A multi-machine multi-product EPQ problem for an imperfect manufacturing system considering utilization and allocation decisions. *Expert Systems with Applications*, *56*, 310–319.

[8] Indrajitsingha, S. K., Raula, P., Samanta, P., Misra, U., & Raju, L. K. (2021). An EOQ Model of Selling-Price-Dependent Demand for Non-Instantaneous Deteriorating Items during the Pandemic COVID-19. *Walailak Journal of Science and Technology (WJST)*, *18*(12), 13398–14.

[9] Cárdenas-Barrón, L. E., Shaikh, A. A., Tiwari, S., & Treviño-Garza, G. (2020). An EOQ inventory model with nonlinear stock dependent holding cost, nonlinear stock dependent demand and trade credit. *Computers & Industrial Engineering*, *139*, 105557.

[10] Ghare, P. M. (1963). A model for an exponentially decaying inventory. *J. ind. Engng*, *14*, 238–243.

[11] Liao, J. J. (2008). An EOQ model with non-instantaneous receipt and exponentially deteriorating items under two-level trade credit. *International Journal of Production Economics*, *113*(2), 852–861.

[12] Tripathi, R. P., & Pandey, H. S. (2013). An EOQ model for deteriorating items with Weibull time-dependent demand rate under trade credits. *International Journal of Information and Management Sciences*, *24*(4), 329–347.

[13] Aggarwal, S. P., & Jaggi, C. K. (1995). Ordering policies of deteriorating items under permissible delay in payments. *Journal of the operational Research Society*, *46*(5), 658–662.

[14] Chen, J. M., & Chen, L. T. (2005). Pricing and production lot-size/scheduling with finite capacity for a deteriorating item over a finite horizon. *Computers & Operations Research*, *32*(11), 2801–2819.

[15] Chang, H. J., Teng, J. T., Ouyang, L. Y., & Dye, C. Y. (2006). Retailer's optimal pricing and lot-sizing policies for deteriorating items with partial backlogging. *European Journal of Operational Research*, *168*(1), 51–64.

[16] Dye, C. Y. (2007). Joint pricing and ordering policy for a deteriorating inventory with partial backlogging. *Omega*, *35*(2), 184–189.

[17] Panda, S., Senapati, S., & Basu, M. (2008). Optimal replenishment policy for perishable seasonal products in a season with ramp-type time dependent demand. *Computers & industrial engineering*, *54*(2), 301–314.

[18] Tsao, Y. C., & Sheen, G. J. (2008). Dynamic pricing, promotion and replenishment policies for a deteriorating item under permissible delay in payments. *Computers & Operations Research*, *35*(11), 3562–3580.

[19] Lin, Y. S., Yu, J. C., & Wang, K. J. (2009). An efficient replenishment model of deteriorating items for a supplier–buyer partnership in hi-tech industry. *Production Planning and Control*, *20*(5), 431–444.

[20] Dye, C. Y., & Ouyang, L. Y. (2011). A particle swarm optimization for solving joint pricing and lot-sizing problem with fluctuating demand

and trade credit financing. *Computers & Industrial Engineering*, 60(1), 127–137.

[21] Wang, K. J., & Lin, Y. S. (2012). Optimal inventory replenishment strategy for deteriorating items in a demand-declining market with the retailer's price manipulation. *Annals of Operations Research*, 201(1), 475–494.

[22] Ghoreishi, M., & Mirzazadeh, A. (2013). Joint optimal pricing and inventory control for deteriorating items under inflation and customer returns. *Journal of Industrial Engineering, 2013*.

[23] Soni, H. N., & Patel, K. A. (2013). Joint pricing and replenishment policies for non-instantaneous deteriorating items with imprecise deterioration free time and credibility constraint. *Computers & Industrial Engineering*, 66(4), 944–951.

[24] Srivastava, M., & Gupta, R. (2014). An EPQ model for deteriorating items with time and price dependent demand under markdown policy. *Opsearch*, 51(1), 148–158.

[25] Wang, C., & Huang, R. (2014). Pricing for seasonal deteriorating products with price-and ramp-type time-dependent demand. *Computers & Industrial Engineering, 77*, 29–34.

[26] Bai, Q. G., Xu, X. H., Chen, M. Y., & Luo, Q. (2015). A two-echelon supply chain coordination for deteriorating item with a multi-variable continuous demand function. *International Journal of Systems Science: Operations & Logistics*, 2(1), 49–62.

[27] Bhunia, A. K., Shaikh, A. A., Sharma, G., & Pareek, S. (2015). A two-storage inventory model for deteriorating items with variable demand and partial backlogging. *Journal of Industrial and Production Engineering*, 32(4), 263–272.

[28] Rabbani, M., Zia, N. P., & Rafiei, H. (2015). Coordinated replenishment and marketing policies for non-instantaneous stock deterioration problem. *Computers & Industrial Engineering, 88*, 49–62.

[29] Indrajitsingha, S. K., Raula, P., Samanta, P., Misra, U., & Raju, L. K. (2021). An EOQ Model of Selling-Price-Dependent Demand for Non-Instantaneous Deteriorating Items during the Pandemic COVID-19. *Walailak Journal of Science and Technology (WJST)*, 18(12), 13398–14.

[30] Indrajit Singha, S. K., Samanta, P. N., & Misra, U. K. (2019). A fuzzy two-warehouse inventory model for single deteriorating item with selling-price-dependent demand and shortage under partial-backlogged condition. *Applications and Applied Mathematics: An International Journal (AAM)*, 14(1), 36.

[31] Rahman, M. S., Duary, A., Shaikh, A. A., & Bhunia, A. K. (2020). An application of parametric approach for interval differential equation in inventory model for deteriorating items with selling-price-dependent demand. *Neural Computing and Applications*, 32(17), 14069–14085.

[32] Kumar, S., Kumar, A., & Jain, M. (2020). Learning effect on an optimal policy for mathematical inventory model for decaying items under preservation technology with the environment of COVID-19 pandemic. *Malaya J Matematik*, 8(4), 1694–1702.

[33] Goyal, A. K., Chauhan, A., & Singh, S. R. (2015). An EOQ inventory model with stock and selling price dependent demand rate, partial backlogging, and variable ordering cost. *International Journal of Agricultural and Statistical Sciences*, *11*(2), 441–447.

[34] Mashud, A. H. M., Hasan, M. R., Daryanto, Y., & Wee, H. M. (2021). A resilient hybrid payment supply chain inventory model for post Covid-19 recovery. *Computers & Industrial Engineering*, *157*, 107249.

Index

For Product Safety Concerns and Information please contact our EU
representative GPSR@taylorandfrancis.com
Taylor & Francis Verlag GmbH, Kaufingerstraße 24, 80331 München, Germany